IET POWER SERIES 14

Series Editors: Prof. A.T. Johns
J.R. Platts
Dr D. Aubrey

Uninterruptible Power Supplies

Other volumes in this series:

Uninterruptible Power Supplies

Edited by John Platts
and John St Aubyn

The Institution of Engineering and Technology

Published by The Institution of Engineering and Technology, London, United Kingdom

First edition © 1992 Peter Peregrinus Ltd
Reprint with new cover © 2007 The Institution of Engineering and Technology

First published 1992
Reprinted 2007

The Institution of Engineering and Technology
Michael Faraday House
Six Hills Way, Stevenage
Herts, SG1 2AY, United Kingdom

www.theiet.org

British Library Cataloguing in Publication Data
A catalogue record for this product is available from the British Library

ISBN (10 digit) 0 86341 263 7
ISBN (13 digit) 978-0-86341-263-9

Printed in the UK by Short Run Press Ltd, Exeter
Reprinted in the UK by Lightning Source UK Ltd, Milton Keynes

Contents

Foreword

'Uninterruptible Power Supplies' is intended to be a fundamental guidebook related to the various systems available and how they may be usefully specified and employed.

While we have come to expect reliability and continuity of supply from the electricity supply utilities, this cannot be guaranteed absolutely and outages and disturbances must be expected at some time. Such outages and disturbances can be quite disastrous in the case of sensitive loads, such as computers, which are now extensively used in modern commerce and industry. Equally, such electronic equipment is often dependent on a clean electricity supply.

Thus, a continuous supply of clean electrical power has become a necessity for many functions within business and manufacturing organisations, and this book outlines what is involved in the design, construction, installation and operation of suitable supply equipment.

After setting the scene in the first two chapters, the next five chapters describe a range of equipments and systems used for uninterruptible power supplies. These are followed by chapters on the application of UPS equipment in active operation in two typical important fields where security of supply is vital.

Finally there are three chapters dealing with, respectively, harmonic distortion, reliability and specification of equipment.

The authors who have contributed to this book represent leading manufacturers, authoritative designers and consulting engineers, and professional engineers of user organisations.

Each author has written his particular chapter based on knowledge and experience gained in day to day involvement with UPS systems. Consequently, the book should be regarded not so much as an academic theoretical treatise, but more as sound engineering logic from experienced practical Chartered Engineers. Therefore, it should prove to be helpful for professional engineers who have a requirement to specify, design, install and subsequently make use of the facilities available from UPS systems.

The Editors wish to acknowledge and give full credit for the initial conception of this book to Alec King and Bill Knight.

The complete team of authors are separately listed; all are expert in their respective aspects and fields of endeavour. Both the IEE and the Editors are grateful for their contributions. Acknowledgment to HM Stationery Office is given in relation to permission to quote extracts from the Electricity Regulations in Chapter 1.

The Editors have tried to produce a unified text, and hope it will fulfil a long felt need.

John Platts
John St Aubyn

Contributors

John Platts
Consultant
Field House
Evesham Rd
Stow-on-the-Wold
Cheltenham
Gloucestershire GL54 1EJ

Bill Knight
Mainstream Electronics Ltd
13–14 The Inner Courtyard
The Whiteway
Cirencester
Gloucestershire GL7 7BA

Gerald Manders
Managing Director
Anton Piller (UK) Ltd
Chesterton Lane
Cirencester
Gloucestershire GL7 1YE

Klaus Sachs
Anton Piller GmbH
Post Fach 1860
D3360 Osterode am Harz
Germany

Brian J Beck
Berymann USA Inc
5701 Main Street
Srite 1100
Houston
Texas 77005–1895
USA

Mike J Leach
Computer Power Division
Emerson Electric UK Ltd
Elgin Drive

Swindon
Wiltshire SN2 6DX

Peter Barnett
Siemens PLC
Sir William Siemens House
Princess Road
Manchester M20 8UR

Ian Harrison
Chloride Industrial Batteries Group
Standby Power Division
Clifton Junction
Swinton
Manchester M27 2LR

Roger Cato
British Airports Authority
Heathrow Airport Ltd
D'Albiac House
Hounslow
Middlesex TW6 1JH

Norman Newton
AYH Partnership Ltd
40 Clifton Street
London EC2 4AY

Alec C King
48 Copthorne Rd
Felbridge
East Grinstead
West Sussex RH19 2NS

Paul Smart
Head of Engineering
Citibank
2 Savoy Court West
The Strand
London WC2

Chapter 1
Why the need arises

J. R. PLATTS

1.1 Reliability of electricity supply

The objective of an electricity supply utility is to provide a wide range and variety of customers with a supply of electrical energy. The purpose is to meet the variable and instantaneous demand for electricity at the most economic cost, and to achieve customer satisfaction by a good standard of reliability and quality, typically in terms of voltage and frequency. Over the last 100 years the customer, whether a householder or a major commercial organisation, has increasingly developed an expectation that electricity will always be available like water from the tap.

In the UK and most other industrial countries total outage of the electricity supply network is an extremely rare occurrence. Reliability is overwhelmingly good except in situations such as very adverse weather conditions. These can be hurricane winds with resultant falling of trees onto overhead distribution lines; flooding due to excessive rainfall with rivers overflowing their banks; and in some countries the build up of ice on overhead distribution lines under conditions known as freezing rain. Freezing fog and rain can produce flashovers on the insulators of overhead lines particularly if they have been previously affected by salt and other pollutants carried by on-shore winds. In economic terms it is totally impossible to safeguard the electricity supply network against sporadic system failures. The majority of disturbances to the continuity of electricity supply are caused by external factors beyond the control of the utility's operating staff. However, through careful design and planning the probability of some disconnections may be reduced, but generally at a price. The balance of costs and benefits of such provision of increased security of supply is a matter for judgment and assessment by the customer.

It is unrealistic to expect the electricity supply utility to plan the generation, transmission and distribution systems such that supplies of electricity are never interrupted. To achieve this ideal would entail investment costs which would make electricity prohibitively expensive, judgments have to be made by the utility related to provision of adequate capacity and the relative trade-off between quality and security of the service for the main body of customers and the costs of providing that service. Although customers have an expectancy that electricity will be available at all times, this availability stems from a myriad of difficult decisions; for example, in the areas of planning, sizing, engineering, control and location of plant and equipment within the composite system.

Most electricity consumers in the United Kingdom experience very rare interruptions in the continuity of electricity supplies. Whilst it is clear that averages can be misleading, the situation is that on average each customer experiences supply interruptions totalling about only one hour 30 minutes per annum (i.e. out of 8760 hours in the year). This average figure has been reduced from one hour 44 minutes per annum applicable in 1965–66, and is inclusive of planned outages. Interruptions are caused not only by adverse weather conditions but also by planned maintenance of distribution networks, of which advance notice is provided by the local distribution company. Needless to say industrial disputes can catastrophically affect outage times.

The Annual Report of Electricité de France indicates that the average outage time for low voltage customers in that country has been reduced from about 11 hours per annum in 1965 to 5 hours 22 minutes in 1986. The figure increased to 6 hours 26 minutes in 1987 due to the effect of the October hurricane. The average outage time due to planned work was 1 hour 15 minutes, due to accidental interruptions was 4 hours 35 minutes and due to generation and transmission difficulties it was 36 minutes.

Outage times of customers in other European countries vary quite considerably. Typically they have been about 18 minutes per annum in West Germany, 38 minutes per annum in Stockholm (Sweden), and 74 minutes for Denmark. Understandably the outage times in rural areas tend to be higher than in urban areas. For Japan it is said that one power company has an average outage time of 6 hours 52 minutes, comprising 46 minutes unplanned and 6 hours 6 minutes for planned work. In the United Kingdom there is a wide variation across the country with average times in the South West and South Wales being about twice as high as in the Midlands, and with London considerably lower than the national average.

1.2 Electricity Supply Regulations 1988

The Electricity Supply Regulations impose requirements regarding the installation and use of electric lines and apparatus of suppliers of electricity including provisions for connections with earth. Part VI contains provisions relating to supply to a consumer's installation, and in particular Regulation 30 imposes a requirement to give information regarding the type and quality of supply with specific limits.

It is equally important to be aware that these Regulations cover provision for 'Discontinuance of supply in certain circumstances' including the causing of undue interference with the supplier's system or with the supply to others. The pertinent extracts are Regulations 27 and 28 which are quoted in this text by kind permission of HM Stationery Office.

1.2.1 *Extracts from Electricity Supply Regulations*

General conditions as to consumers
27(1): No supplier shall be compelled to commence, or, subject to regulation 28, to continue to give a supply to any consumer unless he is reasonably satisfied that each part of the consumer's installation is so constructed, installed, protected and used, so far as is reasonably practicable, as to prevent danger and

not to cause undue interference with the supplier's system or with the supply to others.

27(2): Any consumer's installation which complies with the provision of the Institution of Electrical Engineers Regulations shall be deemed to comply with the requirements of this regulation as to safety.

Discontinuance of supply in certain circumstances

28(1): Where a supplier, after making such examination as the circumstances permit, has reasonable grounds for supposing that a consumer's installation or any part of it, including any supplier's works situated on the consumer's side of the supply terminals, fails to fulfil any relevant requirement of regulation 27, paragraphs (2) to (7) shall apply.

28(2): Where, in an emergency, the supplier is satisfied that immediate action is justified in the interests of safety, he may without prior notice discontinue the supply to the consumer's installation and notice in writing of the disconnection and the reasons for it shall be given to the consumer as soon as is reasonably practicable.

28(3): Subject to paragraph (2), the supplier may, by notice in writing specifying the grounds, require the consumer within such reasonable time as the notice shall specify to comply with one or both of the following:

(*a*) to permit a person duly authorised by the supplier in writing to inspect and test the consumer's installation or any part of it at a reasonable time

(*b*) to take, or desist from such action as may be necessary to correct or avoid undue interference with the supplier's supply or apparatus or with the supply to, or the apparatus of, other consumers.

28(4): In any of the circumstances specified in paragraph (5) the supplier may, on the expiry of the period specified in the notice referred to in paragraph (3), discontinue the supply to the consumer's installation and shall give immediate notice in writing to the consumer of the discontinuance.

28(5): The circumstances referred to in paragraph (4) are:

(*a*) that, after service of a notice under paragraph (3) (*a*), the consumer does not give facilities for inspection or testing; or

(*b*) in any other case:
   (i) after any such test or inspection the person authorised makes a report confirming that the consumer's installation (or any part of it) fails to fulfil any relevant requirement of regulation 27; or

(ii) the consumer fails to show to the reasonable satisfaction of the supplier within the period so required that the matter complained of has been remedied or is the responsibility of the supplier.

28(6): Any difference between the consumer and the supplier in relation to the grounds or the period specified in any notice of the kind mentioned in paragraph (3)(*b*) shall be determined in the manner provided by regulation 29.

28(7): The supplier shall not discontinue the supply in pursuance of paragraph (4) pending the determination of any difference of the kind mentioned in paragraph (6), and shall not discontinue the supply to the whole of the consumer's installation where it is reasonable to disconnect only a portion of that installation in respect of which complaint is made.

28(8): Where in pursuance of this regulation a supplier has disconnected the supply to a consumer's installation (or any part of it) the supplier shall not recommence the supply unless:

(*a*) he is satisfied in respect of the consumer's installation that the relevant requirements of regulation 27 have been fulfilled; or
(*b*) it has been determined in the manner provided by regulation 29 that the supplier is not entitled under regulation 27 to decline to recommence the supply,

and if he is so satisfied or it is so determined, the supplier shall forthwith recommence the supply.

Regulation 30 reads as follows:
Declaration of phases, frequency and voltage at supply terminals
30(1): Before commencing to give a supply to a consumer, the supplier shall declare to the consumer:

(*a*) the number and rotation of phases;
(*b*) the frequency; and
(*c*) the voltage,

at which it proposes to deliver the supply and the extent of the permitted variations of those values:
   Provided that, unless otherwise agreed between the supplier and the consumer, the frequency to be declared shall be 50 Hz and the voltage to be declared in respect of a low voltage supply shall be 240 V between the phase and neutral conductors at the supply terminals.
30(2): For the purposes of this regulation, and unless otherwise agreed by the consumer, the permitted variations are:

(*a*) a variation not exceeding one per cent above or below the declared frequency; and
(*b*) a variation not exceeding six per cent above or below the declared voltage at that frequency where that voltage is below 132 kV, and not exceeding 10% above or below the declared voltage where that voltage is 132 kV or above,

or the variation which may have been authorised by the Secretary of State under paragraph (3).
30(3): The Secretary of State may, on application by a supplier, authorise him to alter any of the declared values or any permitted variation if he gives such notice of his application as the Secretary of State may require.
30(4): The supplier shall forthwith give notice of any authorisation under paragraph (3) to every consumer to whose supply it may apply.
30(5): The supplier shall ensure that, save in exceptional circumstances, any supply he gives complies with the declaration under paragraph (1).
30(6): The polarity of direct current and the number and rotation of phases in any supply shall not be varied without the agreement of the consumer or, in the absence of such agreement, the consent of the Secretary of State who may impose such conditions, if any, as he thinks appropriate.

Figure 1.1 *The Halifax Building Society's Copley data centre is protected by an Anton Pillar uninterruptible power supply system backed by over 3000 Tungstone ultra high performance Planté cells. Power can be supplied for up to 20 min to allow a back-up supply from diesel generating sets to be brought on-line or, if the generators fail to start, for a controlled shut down of the computers to be completed. The data centre contains computers which control over 10 000 different terminals and automatic teller machines and are permanently on-line. The UPS systems are tested weekly*

1.3 The need to protect against power supply breaks and disturbances

Whilst the general standards of supply are completely adequate for most applications in industry and commerce, the contamination of the supply by relatively small disturbances can cause serious and unacceptable problems for sensitive loads — typically computers. The need is not just to provide standby power in the event of supply failure, but is also to make certain that the electrical input to valuable computer systems is as pure, clean and continuous as is required, to prevent any data or control signals being corrupted or lost entirely (Fig. 1.1).

Although from the information already provided it can be seen that total power outage from the electricity supply utility is infrequent, it should be recognised that the timing of a break is beyond the customer's control. It follows that a sudden break in power supply or even a mild fluctuation could occur at a critical time for the business. This might prove to be disastrous, since the computer's data could become corrupted or totally lost. The manifestation of the incident may become apparent through damage to computer hardware as well as sporadic and unexplainable errors occurring on the software.

Similarly, mains-borne electrical interference such as surges can adversely affect computers and process control equipment. Typically this electrical interference results from thunderstorms and lightning, from switching operations and from other electrical apparatus within the building (e.g. from thyristors within electronic devices utilised for controlling other essential electrical equipment).

One might have expected the computer and process control manufacturers to have designed their equipment to be less susceptible to electrical interference arising from spikes, dips and surges. Apparently, this has not proved to be possible, and for this reason there has been progressive development of uninterruptible power supply (UPS) systems. They become an economic proposition for organisations whose business and associated information technology demand a vital need to protect against power supply breaks and disturbances.

1.4 Electrical interference and contamination

Several forms of interference can cause problems for electronic instrumentation and information technology systems. Each form of electrical interference is separately considered in this section as follows:

1.4.1 *Spikes*

Lightning is the atmospheric condition which is the main external and unpredictable causes of spikes on the public electricity supply network. Lightning strikes which hit overhead lines can produce high voltages spikes of some considerable power. Also similar effects are the result of ground strikes of lightning by induction into power lines and cables.

Other major sources of spikes are switching operations and particularly those relatively close to the computer and other electronic equipment. Switchgear operating on heavy currents and HRC fuses are devices which are known to produce transient spikes when operating as designed and as intended under fault conditions.

Spikes with high amplitudes up to many hundreds of volts, and even in excess of 2 kV, are to be encountered with major electrical system faults. Being transient, they are of short duration and last for some 100 µs. They are oscillatory having a frequency in excess of 100 kHz. The rise time can be relatively short at about 1 µs.

A greater occurrence of spikes is attributed to semiconductor switches, and they can even be caused by electrostatic discharges from staff walking on office carpeting and subsequently touching metallic cabinets with a resultant spark. In particular, repetitive strikes are generated by thyristors and triacs used in control systems; by variable speed commutator motors as used on lifts and escalator drives; by sodium and mercury discharge lamps; by mercury arc rectifiers in various applications for the production of direct current; and also by the overhead collector or third rail collector on electric traction systems. Repetitive strikes attributable to semiconductor devices have amplitudes normally not exceeding 300 V and with a rise time in the range of 50 ns to 1 µs. Total duration time does not normally exceed a few tens of microseconds.

Other forms of spikes are generated by the switch starting and stopping of induction motors, which are typically used for driving pumps and fans. Random spikes commonly have amplitudes up to about 800 V and rise times as short as a few nanoseconds. Contact bounce of the contractors within the starter can result in a series of spikes; although each spike has a duration of less than 100 μs, the contact bounce will extend the incident over a longer period.

1.4.2 *Surges and dips*

Surges and dips in the electrical supply are generally brought about by the switching off (surges) and switching on (dips) of large loads. The criterion is the size of the load being switched in relation to the transformer rating at the local point of supply. The phenomenon of a surge is analogous to the momentum of inertia in a mechanical system, and a dip is a reverse concept to a surge. Typically, one can sense and become aware of surges and dips through variations in the luminance of lamps in lighting systems. Duration periods for both surges and dips can be in the range of a single half cycle of the electricity supply waveform through to several half cycles. The latter is the more usual duration period. The variation from the nominal voltage resulting from surges and dips can exceed the normal range of 6% above or below.

Types of loads to be switched within the building and fed from the same electricity supply point are usually electric motors. These may be as applied for motive power to drive lifts, escalators, refrigeration compressors, pumps or fans.

Beyond the local point of supply, voltage dips are sometimes caused by major variations on the high voltage supply network such as system faults or sometimes major load switching of industrial plant. Normally provision is made at the design stage to avert the adverse impact of industrial plant such as arc furnaces. The number and size of voltage dips due to disturbances beyond the point of supply is to a large extent dependent upon the robustness of the local network. Generally, the situation prevailing in a major city or urban area is better than in a rural area.

Fluctuating high power loads are encountered in industrial processes. For example, in steel works variable high power loads could arise from arc furnaces and from highly powered reversing mill motor drives. These types of installation can produce voltage fluctuations which, in transmission through to the supply network, may result in flicker. The flicker usually has frequency of a few hertz with voltage amplitudes within a few percent of the nominal voltage.

1.4.3 *Harmonics*

Thyristors, mercury arc rectifiers, discharge lamps, diodes, induction motors and variable speed commutator motors are all electrical devices which generate harmonics on the mains supply.

These may be relatively low on most distribution systems with the total harmonic content of the voltage in the range of 2—5%. Only under exceptional circumstances will higher total harmonic contents be experienced.

1.5 Criteria influencing the need for UPS

Uninterruptible power supplies are necessary for any business operation which requires a very high availability factor and purity factor for a key facility on which the core activities crucially depend. For applications where corruption of data or interruption of supply even for a fraction of a second cannot be tolerated, a UPS is needed.

A surge, dip, break, fluctuation, or other contamination can in some situations prove to be absolutely devastating. Such circumstances can result in disastrous loss of data. The most susceptible and vulnerable facilities are the whole range of computer and instrumentation processes.

The vulnerability of a computer system or information technology system to interference or mains supply disturbance is dependent upon the precautions built into the system by the designer.

Chapter 2

History and market growth

W. R. KNIGHT

2.1 Early development of UPS equipments

The need for reliability in power supplies has obviously been with us for many years. With the increased use of electrical power and our dependence on an electrical supply, reliability has become an increasing concern. Initially the market for uninterruptible power supply equipments was of low demand and was led by users who sought clean and secure electricity supplies from their various engineering groups.

The first uninterruptible power supply equipments (then known as no-break power supplies) were of rotary design, and so far as the author is aware these appeared during the 1950s. The market at that time was related to defence equipment such as communications and radar. Although many arrangements were tried, the most popular was that shown in Fig. 2.1. Earlier sets were probably developed for military purposes during the 1939–45 war and the development continued thereafter to achieve greater reliability, increase in efficiency and reduction of maintenance.

Referring to Fig. 2.1, normal operation was for power to be supplied through the rectifier to the DC motor which in turn powered the AC generator. On loss

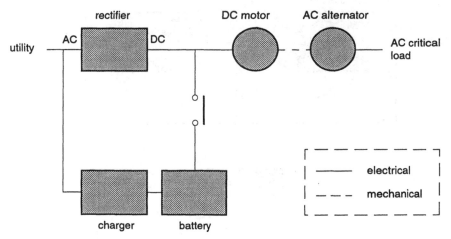

Figure 2.1   *DC motor alternator with battery back-up*

of mains the battery became the energy source, and enabled the generator to supply uninterrupted power. Since the battery voltage would have dropped from say 2·25 to 1·9 V per cell, the flywheel was used as a transient power source to ensure good regulation during the operation of the comparatively slow DC speed regulator.

Early power rectifiers were, in general, either selenium or mercury arc, and for UPS purposes selenium would have been the obvious choice. Silicon and germanium rectifiers became available in the early 1960s and these were more efficient and used less space. A few years later the advent of the thyristor (then known as the silicon controlled rectifier) allowed the development of static inversion equipment; previously inversion had required the use of the mercury arc rectifier. These two developments led to the appearance of static UPS equipment.

The first thyristor suffered occasional failures but over the years the physics of junction devices became better understood, and manufacturers were able to introduce improved techniques and quality control. With time, the static systems achieved the pre-eminent position of the rotary systems.

The advantages claimed for the static systems were:

(i)   Higher efficiency: 85–91% against 84–88% for rotary
(ii) Compactness: reductions in size of some 20%
(iii) Lower maintenance costs
(iv) No brushgear (varying loads involved changing the number of brushes to maintain current density)
(v)   Lower noise level; e.g. a reduction from 94 dBA down to 72 dBA
(vi) Higher reliability

Figs. 2.2, 2.3 and 2.4 depict the operation of the static type of UPS equipments: Normal operation is shown in Fig. 2.2; power is supplied through the rectifier/

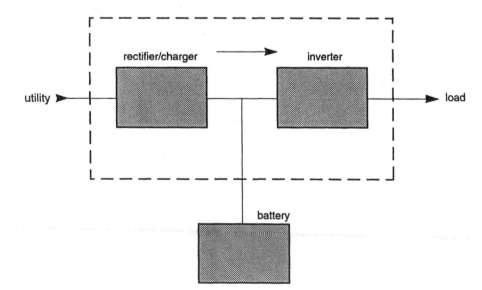

Figure 2.2   *UPS normal operating mode*

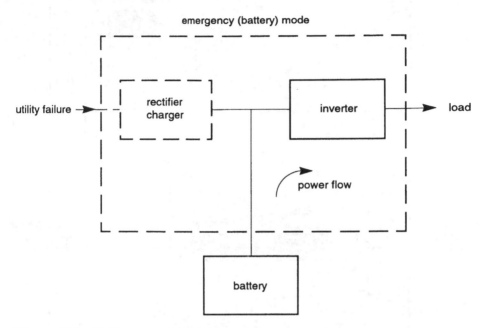

Figure 2.3 *UPS operation with failed utility power*

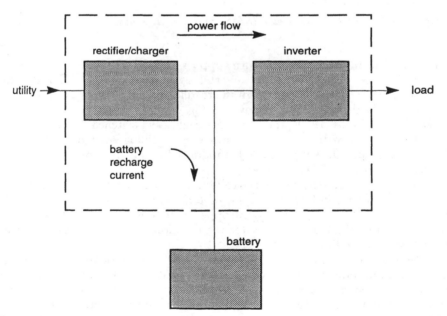

Figure 2.4 *UPS operation after return of utility power*

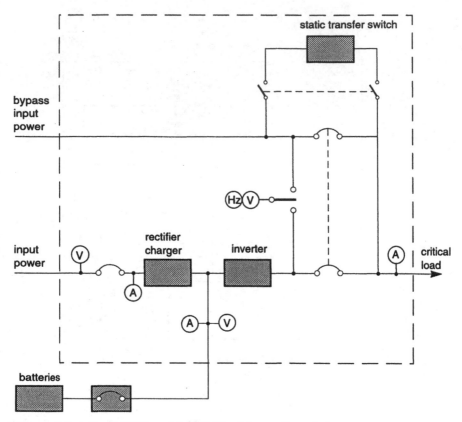

Figure 2.5   *Single UPS module (SMS) with transfer switch*

charger to the inverter which in turn supplies the load. On loss of mains, power is supplied from the battery to the inverter, as shown in Fig. 2.3. On restoration of mains, power is supplied through the rectifier/charger to recharge the battery and to supply the inverter, as shown in Fig. 2.4.

During the mid 1970s the use of a statically switched by-pass became possible, thus providing a smooth uninterrupted changeover to raw mains if a UPS fault occurred. A typical single module UPS system, complete with static by-pass is shown in Fig. 2.5.

In addition to increasing the reliability, the static by-pass solved the voltage drop problem caused by load surges, motor starting, or fuse rupture on faults. Whenever the UPS output voltage dropped, for whatever reason, the load could be switched to the mains supply for the duration of the surge, and thereafter returned to the UPS.

Where the reliability of a single module UPS is considered inadequate, a parallel redundant system may be used as indicated in Fig. 2.6. The load capability of such a system is $(N-1) \times P$, where N is the total number of modules installed and P is their rating. The loss of any one unit does not prevent the system from remaining in service.

Static and rotary UPS systems have developed alongside each other, as will be apparent from Chapters 3—6. The electromechanical relays of the early rotary systems were replaced by solid state logic systems, and sophisticated techniques and controls were introduced. New techniques continue to be introduced and the variety of UPS arrangements, based on combinations of rectifiers, inverters, alternators and diesel engines, continues to increase.

The later designs of static systems tend to use pulse width modulated inverters with either transistor or thyristor switching. The higher frequencies of switching (up to 40 kHz for transistors) reduce the size and weight of the

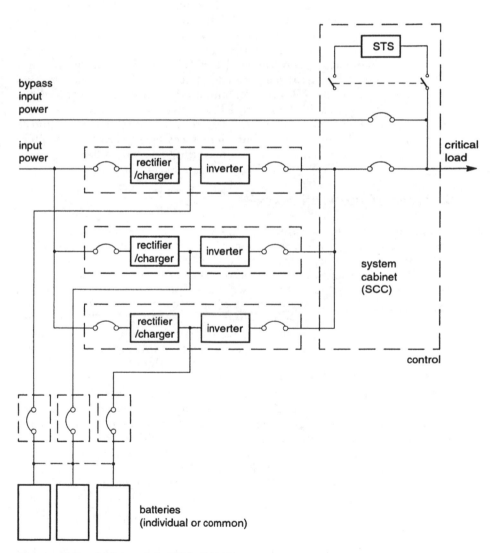

Figure 2.6   *Multi-module UPS (MMS) with system control cabinet (SCC)*

inductive components. The forced commutation required for thyristors produces acoustic noise; transistors are switched and do not require forced commutation; less noise is therefore produced.

2.2 Later developments of UPS equipments

Systems are developing along the lines of achieving higher efficiency, less space utilisation, ease of installation, and user-friendly interfaces. Instead of being presented with a multiplicity of switches and analogue meters, the operator sees a few push buttons and indicator lamps. Measurements are displayed digitally, and the run-up procedure is probably automatic. Operation is becoming simpler, and the equipment is therefore becoming more reliable.

Now that sealed recombination lead–acid batteries are the normal choice, the opportunity is frequently taken to install the UPS equipment in the computer room, rather than in a distant plant room. Some computer manufacturers are tending to incorporate the UPS equipment into their own enclosures so that it becomes, in effect, a part of the Central Processing Unit (CPU).

Servicing and maintenance may be carried out by undertaking a diagnostic check using a small built-in or portable microprocessor. In most cases of failure, the service technician merely has to replace printed circuit boards.

2.3 Areas of the UPS market

Although the major share of the market is with computers and communication equipment, there are other growing market areas. One is in factory automated production where numerically controlled machinery requires high quality power. The installation of a UPS ensures continuity of power and eliminates interference caused by the continual switching of heavy machinery. The use of a UPS for this application is growing, and there certainly is now appearing a market for equipments to operate whole segments of a production line. In a works environment the diesel engine driven rotary system, which can also act as a peak lopping device, has a significant advantage over the static system, but at the lower ratings (say 200–300 kVA), the static system has the advantage of being less obtrusive and more compact (Fig. 2.7).

Another growing area is the supply of UPS for desk mounted computers (PCs) or word processors (Fig. 2.8). An increasing number of such computers is used in continuous process or in the quick and easy retrieval of documents from a memory bank, for example in a doctor's surgery. Market forces have induced the production of a very simple UPS which is compatible with, and fits neatly under, the minicomputer, being only 60 mm deep.

At low system ratings (up to 10 kVA) the system may be either on-line, off-line or triport.

For on-line systems the load is normally supplied through the UPS and uses raw mains only when the by-pass is in operation.

For off-line systems the load is normally supplied from the raw mains and uses the UPS system only during a mains failure or during disturbances.

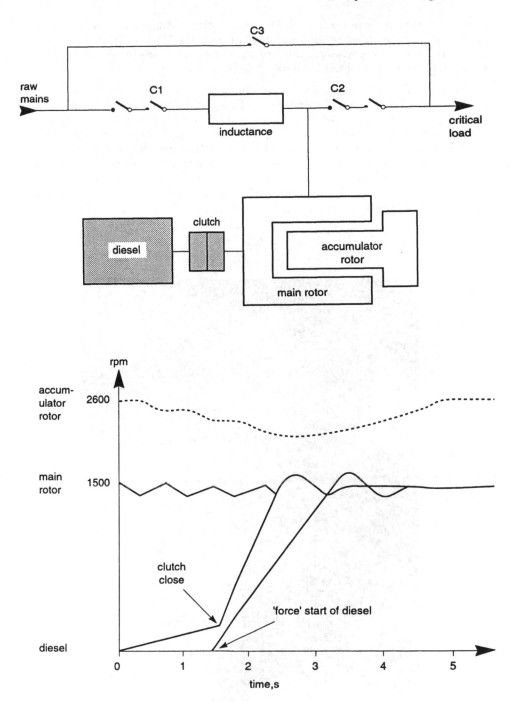

Figure 2.7 *Function of KS stato-alternator*

The triport system incorporates a solid state switching arrangement which may be used as either a rectifier or an inverter. Normally the load is supplied from the mains and the rectifier charges the battery. On loss of mains the rectifier is used as an inverter and the load is supplied from the battery.

Equipment used in the off-line mode need not be continuously rated and is therefore smaller, simpler and less costly. Efficiency of the system is higher since the inverter is not normally in operation. The load may be protected against mains borne interference by a filter network and, in the USA, off-line units up to 200 kVA are currently available. The author believes that the market will certainly show an increasing demand for such systems, with the advent of a quick response time from the static switch and improved filter performance. It seems that the majority of the market could end up with this design, particularly at low VA rating levels; the advantage of the reduction in losses is significant.

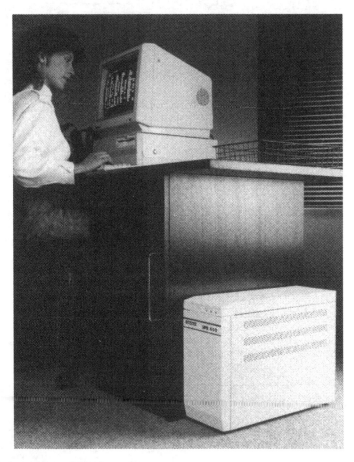

Figure 2.8   *Simple UPS for use with microprocessor*

2.4 Market growth

The present size of the overall UK market is approximately £70 million per annum. Generally overall market growth is 15–17% per annum with a diminution in growth rate at higher system ratings. Systems above 50 kVA show only a small percentage increase whilst the market growth at the bottom end (say up to 2 kVA) is high and is thought to be 30–35% per annum.

The market for equipments between 2 and 50 kVA is also growing at a fast rate, and the author expects to see this growth continue for the next five to ten years. It is also felt that the rotary set will continue to take an increasing portion of the market at higher ratings. For example, the latest KS rotary system, with high efficiency (97%), compactness and exceptionally good electrical characteristics, has advantages over the static systems above 400 kVA.

2.5 Future trends

It is always difficult to predict future changes, but the following suggestions would seem to be probable.

Future designs for large computer centres will include a combined suite of cubicles for air conditioning, raw power distribution and high quality power distribution. This will produce obvious savings in space and reduced maintenance costs.

The market for high reliability power for continuous manufacturing processes will tend to increase. We are likely to see factories in which power distribution, UPS equipment and prime movers are located in one room or area. The UPS will merge into the scheme of factory power supply and distribution.

In the office environment there will be an increased tendency for complete buildings, designed around the modern concept of computers and communications, to be built as standard units for any operating client to lease or rent. Such speculative projects will be designed and built to incorporate all services needed, including air conditioning, false floors, special lighting and high reliability power.

Chapter 3
Dynamic systems with battery energy store

G. MANDERS and K. SACHS

3.1 Introduction

As outlined in Chapter 1, the reasons for using a UPS system are well known: mains disturbances of all kinds, from short-circuits to prolonged interruptions, frequency and voltage fluctuations, affect sensitive, important loads. The UPS system provides a better quality voltage to the load. It is obvious that the operation of the UPS itself must not be disturbed by distortion and spikes on the mains supply.

Recently, however, problems of non-linearities of the mains supply have become increasingly important. Voltage distortion is caused by harmonic currents in the input rectifier. In most countries there are recommendations, in a few also laws, which lay down limits for this distortion. The design of a modern UPS must take these into consideration.

Disturbances occur on the output side, which are caused by the load itself. These are interference from non-linear currents in switched-mode power supplies, overloads due to switching on loads, short-circuits and the special dynamic characteristics of controlled loads. UPS operation must not be affected by these and the voltage waveform should be distorted as little as possible by the suitably low-impedance output of the UPS.

3.2 Uniblock convertor

The above-mentioned conditions are largely fulfilled by UPS systems with rotating convertors. The heart of the UPS concept demonstrated here is the Uniblock convertor. This is a synchronous machine in which the motor and generator windings are both located on a single, common stator. The rotor is DC excited and carries a damper winding similar to the cage of an asynchronous machine (Fig. 3.1). This form of construction produces a low unit volume with relatively short distance between bearings.

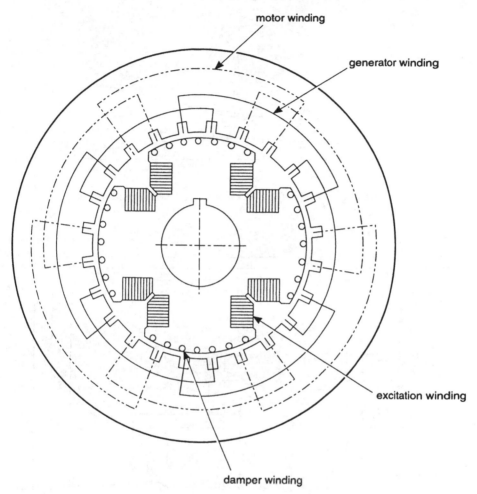

Figure 3.1 *Uniblock convertor: arrangement of windings*

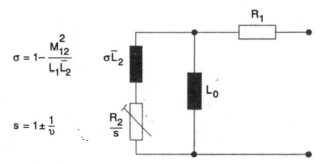

Figure 3.2 *Equivalent circuit diagram for an asynchronous motor*

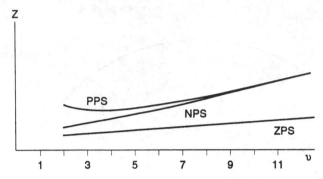

Figure 3.3   *Frequency versus output impedance for an asynchronous motor*

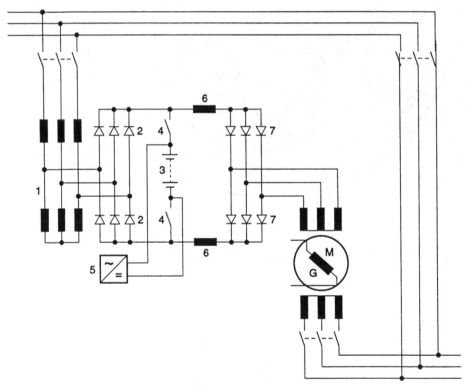

Figure 3.4   *UPS system with Uniblock convertor*

1   Matching transformer
2   Mains rectifier
3   Battery
4   Battery switch
5   Charger
6   Smoothing choke
7   Inverter

In this machine the transfer of energy from the motor to the generator is mainly due to the alternating magnetic flux produced by the motor winding, to which both windings are exposed. This alternating flux circulates as a rotating field and drives the rotor synchronously. From the rotor's point of view the rotating field appears as a unidirectional flux which can be influenced by additional DC excitation of the rotor. This allows the amplitude of the rotating field to be controlled independently of the input voltage and thus, by suitable regulation, maintain the voltage at the generator winding constant. The energy transfer to the rotor is by the usual brushless method.

Harmonics in the input voltage generate higher-frequency rotating fields, for which the damper cage rotates asynchronously and represents a short-circuit. Therefore, for the suppression of harmonics the equivalent circuit diagram for asynchronous motors shown in Fig. 3.2 applies. As there are only resistors and inductors, resonance cannot occur. The frequency response of the internal impedance is therefore expected to have a characteristic without irregularities or poles. By suitable design of the machine, the internal resistances and inductances or their mutual couplings, respectively, can be greatly influenced, thus achieving low effective impedances.

Harmonics create fields of higher order which can rotate in positive (PPS) or in negative phase sequence (NPS). The direction of rotation is expressed by the polarity of the term $1/v$ in the expression for the slip s, where the positive sign applies to a counter-rotating field. This results in two curves for the output-impedance/frequency which have to be taken into consideration (Fig. 3.3).

In addition to the rotating field systems, zero phase systems (ZPS) will occur, mainly for harmonics with ordinates that can be divided by three. These appear in phase in the three-phase windings and cancel each other out in whole or in part if the windings are suitably pitched. A third impedance curve applies for this, which generally lies much below that applicable to rotating field systems. The transmission of harmonics to the generator side is therefore effectively suppressed. For current harmonics generated on the load side, the damper cage appears as a short-circuit in parallel with the generator winding. The output impedance — and thus the effect of the harmonic currents on the output voltage — is considerably reduced by this method.

Since two windings of rated output, namely a motor and a generator winding, have to be fitted to the stator, the kVA rating of the Uniblock corresponds to twice that of a standard synchronous generator of the same rated output. This results in winding reactances which are half the value of those in a machine with the conventional form of construction.

If the supply voltage is removed, the rotor maintains the magnetic flux in the machine. The energy taken from the generator winding is then extracted from the rotating masses.

Alternatively, this machine can also be considered as a three-phase transformer with flux control, which only transmits the fundamental component and represents both an energy storage mechanism and a source of reactive current. Although this machine is somewhat more difficult to produce and has a higher overall weight than a comparable three-phase transformer, it requires the same amount of winding copper and core iron.

Anton Piller have coined the term Uniblock convertor as the motor and generator are combined together in a single machine.

3.3 UPS systems with Uniblock convertors

Fig. 3.4 shows a standard UPS system with Uniblock convertor. During mains operation the existing mains voltage is converted into a DC voltage by an uncontrolled rectifier (diodes). At low power ratings this operates as a 6-pulse three-phase bridge circuit. At higher powers the necessary matching transformer feeds a 12-pulse rectifier and thereby reduces the system perturbation.

During mains failure the battery is connected and assures a continuous supply of power to the DC link circuit. The brief zero current break is bridged by the rotating masses. The DC voltage is reconverted to a three-phase AC voltage in a thyristor convertor. This convertor operates like a line-commutated inverter where the active power required to run the motor is taken from the DC circuit. The reactive control and commutation power is provided by the synchronous motor.

Since the motor is run at constant speed and constant voltage, the convertor can be triggered directly from the terminal voltage of the motor. The active current up-take of the motor, and thus its torque, can be controlled by controlling the turn-on phase of the thyristors in relation to the AC voltage. With constant motor voltage and speed, this is more or less proportional to the current flowing in the DC link circuit. Speed control can therefore be provided by a speed control circuit with secondary current control loop, as is usual with thyristor-fed DC drives. The current control circuit provides quick-acting current limiting to protect the entire system, and generator load current feedforward control.

Since at standstill the synchronous motor cannot produce the necessary commutation reactive power for the convertor, a pony motor is provided for starting. Triggering of the convertor is inhibited during starting until the synchronous motor has reached its rated voltage and approximate rated speed.

The generator winding feeds the critical loads in the usual way. A by-pass switch, which is also activated during malfunctions, is provided for servicing operations.

The set can also be operated without a battery. In this case the energy stored in the rotating masses effectively bridges brief interruptions up to 100 ms, i.e. about 90% of all breaks. Later expansion to a battery-backed UPS system is easily achieved.

3.4 Systems with internal redundancy

The simple construction of the Uniblock machine results in an extremely high reliability. The MTBF is greater than 10^6 h. With redundant systems it is not therefore necessary to include the convertor machine in the redundancy arrangement. It is only necessary to build redundancy into the components of the power circuits and control electronics such as rectifiers, convertors, switching and control devices.

This method enables considerable space and weight savings to be achieved, as well as reduction in cost. Fig. 3.5 show such a redundant arrangement. Apart from the power supply for the synchronous motor of the Uniblock convertor, consisting mainly of the rectifier 2 (or alternatively, battery 3) and the inverter

7, a further infeed is provided via a thyristor switch 8 and line reactors 9. In normal operation energy is principally supplied via the latter circuit formed by the thyristor switch and AC reactors. The inverter 7 is controlled so that it contributes only a small proportion of the power. In this way its functions can be monitored continuously without large power losses occurring due to double energy conversion from three-phase AC to DC current and vice versa.

The thyristor switch, whose thyristors are monitored continuously, is controlled so that active power flows only in the direction of the motor. Reactive power, however, can be exchanged in both directions. In the event of a mains failure, this prevents energy from flowing into loads in parallel with the mains or into a short-circuit. The energy stored in the flywheel masses is thus retained. The reactors 9 limit the reactive current and facilitate trouble-free operation even when there is a large difference between the motor voltage and the mains voltage.

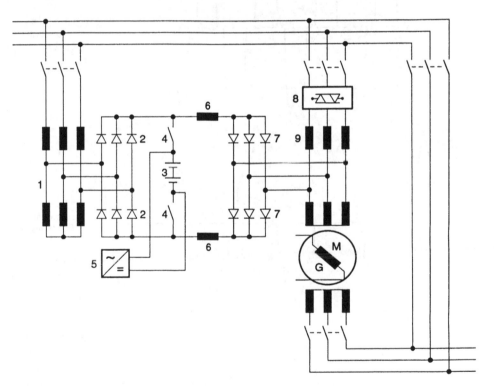

Figure 3.5 *UPS system with Uniblock convertor and internal redundancy*

1 Matching transformer
2 Mains rectifier
3 Battery
4 Battery switch
5 Charger
6 Smoothing choke
7 Inverter
8 Thyristor switch
9 Line reactor

The protected busbar gets its power initially from the flywheel masses of the rotating systems and from the battery after the battery switch 4 has closed. During malfunctions in the rectifier and inverter, this branch is mechanically isolated and the motor is fed exclusively via the thyristor switch. If the rectifier fails, battery operation is still possible; if the inverter fails the system has stabiliser characteristics, i.e. transient breaks of up to 100 ms are still overridden. If the thyristor switch fails, this circuit is mechanically isolated and the motor is supplied via the rectifier 2 and the inverter 7.

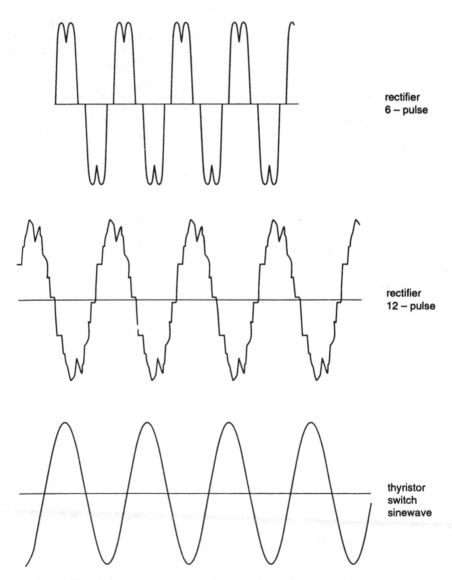

rectifier
6 – pulse

rectifier
12 – pulse

thyristor
switch
sinewave

Figure 3.6   *Uniblock convertor: input currents in the various types of system*

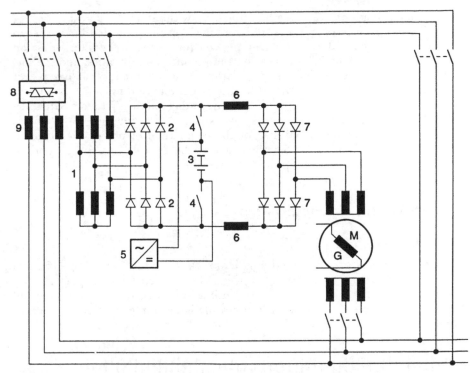

Figure 3.7 *UPS system with Uniblock convertor: thyristor switch connected to generator side*

1 Matching transformer
2 Mains rectifier
3 Battery
4 Battery switch
5 Charger
6 Smoothing choke
7 Inverter
8 Thyristor switch
9 Line reactor

In the event of excessive frequency deviation in the mains, the feed is also via the rectifier/inverter, whereas the thyristor switch is turned off. After return to the permissible frequency range, the thyristor switch is automatically connected during the in-phase condition. Owing to the two redundant branches, formed by the rectifier/inverter or thyristor switch, respectively, an MTBF figure of over 600 000 h is achieved for the whole system.

Apart from the elimination of losses by avoiding the double energy conversion in the DC link circuit, another advantage of this system concept is the freedom from mains disturbances as a result of the near sinusoidal input current. Fig. 3.6 compares the input current waveforms when operating with 6-pulse and 12-pulse input rectifiers and with a thyristor switch.

A further reduction in losses can be achieved if the by-pass branch is connected to the generator side of the Uniblock convertor, as the convertor is then run off load as a reversible machine without electrical isolation. A generally acceptable voltage quality is obtained on the protected busbar, even with mains voltage with severe harmonic content, owing to the filtering action of the line reactors which present a high impedance to harmonics, and the low impedance of the synchronous generator. Harmonics and reactive power generated at the load are absorbed by the generator winding and this relieves the mains to a considerable degree. When operating via the thyristor switch, i.e. normal operation, the current in this case is also largely sinusoidal. Efficiencies of over 95% are achieved because of the elimination of practically all load-dependent losses (Fig. 3.7). Several sets can be operated in the redundant parallel mode to increase the power rating.

3.5 Operating characteristics

3.5.1 *Overload*

Because of the masses in the windings, Uniblock convertors have a high thermal inertia. They can therefore be substantially overloaded for short periods (few minutes) without being damaged. To utilise this high overload capacity all semiconductor devices in the circuit are designed for a continuous current which is 1.5 times the rated current.

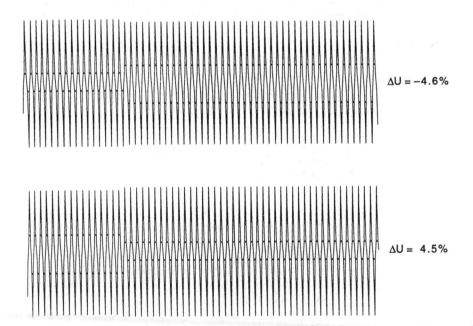

ΔU = −4.6%

ΔU = 4.5%

Connecting and disconnecting 50% load
PF = 0.8 (50% previous load)
UNIBLOCK 300 kVA, 480 V, 60 Hz

Figure 3.8 *Uniblock convertor: performance during load variation*

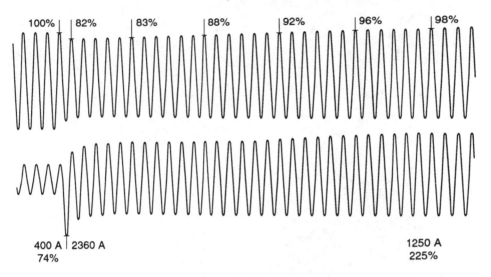

Connecting a 90 kW asynchronous motor
UNIBLOCK 450 kVA, 460 V, 60 Hz, 540 A

Figure 3.9   *Uniblock convertor: performance during surge load*

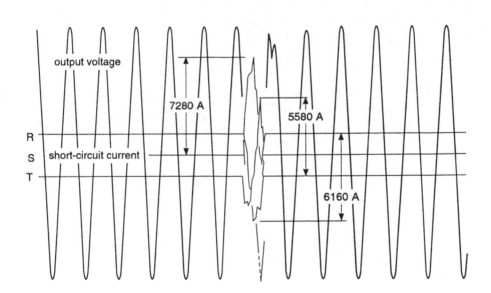

3-phase short-circuit, 240 A fuses
UNIBLOCK 300 kVA, 415 V, 50 Hz, 417 A

Figure 3.10   *Uniblock convertor: performance under short-circuit conditions*

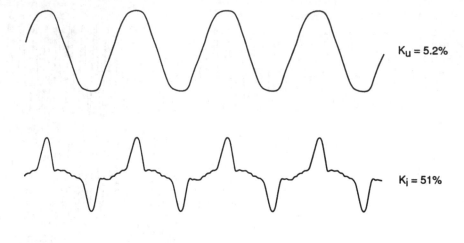

$K_u = 5.2\%$

$K_i = 51\%$

$I_{eff} = 123\ A \qquad I_{spitze} = 272\ A \qquad crest\ factor=2.2$

UNIBLOCK 200 kVA, 400 V, 50Hz

Figure 3.11   *Uniblock convertor: performance with non-linear load*

3.5.2 *Changes in load*

When loads are connected or disconnected the internal voltage drops appear as voltage dips or overvoltages. The voltage regulator then acts to correct the excitation to return the output voltage to the rated value. Owing to the low

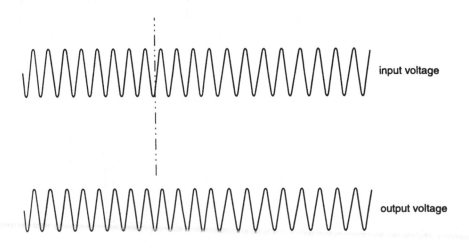

input voltage

output voltage

capacitor discharge 50 μF, 2 kV
UNIBLOCK 70 kVA, 380 V, 50Hz

Figure 3.12   *Uniblock convertor: performance under mains overvoltage conditions*

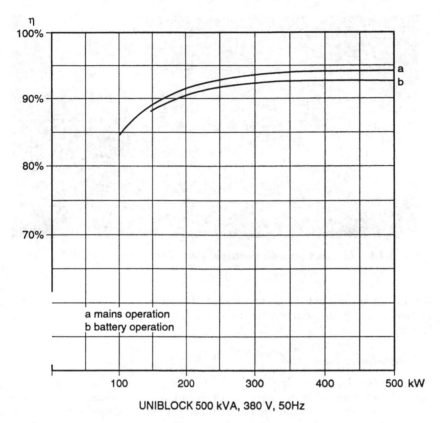

Figure 3.13 *Uniblock convertor: overall efficiency of a UPS set*

internal resistance, these voltage dips are unusually small. Values of ±5% when switching at half load are typical for Uniblock machines, as Fig. 3.8 shows.

3.5.3 *Surge loads*

The performance is also satisfactory under surge load conditions. Fig. 3.9 shows the performance of a Uniblock convertor when directly connecting an asynchronous motor at an initial loading of 3/4 of rated load. In spite of a starting current of more than twice the rated current, the machine is able to maintain the terminal voltage at the end of the recovery period. The entire starting cycle lasts for only a few seconds.

3.5.4 *Short-circuit*

An important property of a secure power supply is its ability to selectively shut down faulty loads as quickly as possible. The majority of critical loads tolerate transient interruptions of only 10 ms maximum. Owing to its low internal impedance, the Uniblock convertor develops short-circuit currents of about ten

Figure 3.14   *Uniblock convertor: internal view of series up to 600 kVA*

times the rated current. Even heavy-duty fuses can be disconnected within the specified time by this current without recourse to the primary network. (Fig. 3.10).

3.5.5 *Non-linear loads*

Owing to the low impedance of the winding the voltage drop across the generator winding due to harmonics is small. This means a low voltage distortion factor without the use of filters, even with severely distorted currents.

Fig. 3.11 shows voltage and current waveforms of a UPS operating with a non-linear load at the data centre of a major UK clearing bank. Despite a current distortion factor of over 50%, with a 50% 3rd-harmonic component, the voltage distortion factor is only 5%. If the individual harmonics of voltage and current are taken together and the load-current/rated-current ratio of the machine is taken into account, a short-circuit reactance of 5.8% is obtained. This figure more or less corresponds to that of a conventional high-voltage transformer in the supply to a building.

3.5.6 *Interference suppression*

The Uniblock machine very successfully suppresses noise voltages at the input owing to the interaction of chokes, stator and damper windings, winding inductance and capacitance. Even fast voltage transients are attenuated in the machine to such an extent that they are no longer perceivable at the output. Fig. 3.12 shows the glitch due to a capacitor discharging, with an energy content of 100 joules.

3.5.7 *Efficiency*

The efficiency of a Uniblock convertor is exceptionally high because of the elimination of electromechanical energy conversion. Unusually high efficiencies of well over 90% can be obtained for UPS systems — especially when operating directly from the mains via the thyristor switch, where the losses in the rectifiers and inverters are eliminated as well. Fig. 3.13 shows the efficiency graph of a 500 kVA set under load. Here a maximum efficiency of over 94% is achieved. The satisfactory shape of the efficiency curve in the partial-load range should also be noted.

3.5.8 *Noise level*

The Uniblock convertor is housed in a common cubicle along with the associated switchgear and control devices. The cooling air required by the machine is produced by a fan which also cools the other equipment in the cubicle. Enclosure in the cubicle efficiently reduces machine and fan noise. An additional silencer can be fitted if required. Fig. 3.14 shows the internal view of a UPS set from the series up to 600 kVA. The Uniblock machine can be seen in the centre with the exhaust air duct towards the top. Above it are the control electronics and circuit breakers. The power electronics are located in the left-hand part of the cubicle.

Chapter 4

Rotary systems with integral diesel engine

B. J. BECK

4.1 Introduction

Within the family of rotary UPS systems are those which, instead of using battery banks as the short term energy store, use a form of kinetic energy storage and an integral diesel engine. These systems provide uninterrupted or no-break power, utilising the kinetic energy store for a matter of seconds and the diesel engine for long term energy supply.

The rotary solution carries the associated advantages of rotating machinery; the generation of a sinusoidal waveform and the ability to provide short-circuit currents being two dominant facets. A diesel UPS system can be considered as a power generating station located adjacent to the critical load. This can represent a subtle difference to other types of system which may be broken into modules and distributed throughout an installation.

Diesel UPS systems divide into two main areas:

- Fixed flywheel design
- Induction coupling design

4.2 Fixed flywheel

Fig. 4.1 indicates the general arrangement of such a system. The system stores kinetic energy in its flywheel during periods that the mains are healthy. The critical load is supplied by the alternator which is driven by the electric motor.

Under mains failure conditions the kinetic energy stored in the flywheel drives the alternator and hence maintains the load to the consumer. Simultaneously the diesel engine will be started via the electromagnetic clutch, run up to speed and take over the driving force to the alternator.

This type of system is of a rugged construction and has been the subject of a number of recent improvements:

(a) *Frequency combinations*: A gear unit between the motor and alternator enables the system to be operated with a 50 Hz input supply and a 60 or 400 Hz output.
(b) *Shortbreak*: The alternator of the system supplies true no-break power. However, for those parts of a consumer's installation that can withstand a short-term interruption, the synchronous motor operates as an alternator after a period which is selectable between 0·2 s and 10 s.

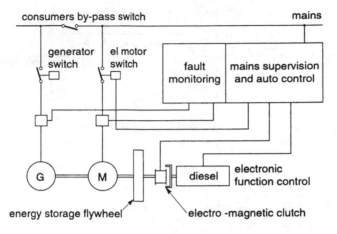

Figure 4.1 *Fixed flywheel system*

(c) *Static convertor/synchronous-motor (Fig. 4.2)*: The inverter is connected in series with the synchronous machine and thus can maintain a close-tolerance output frequency to the motor whilst accepting a mains input which exhibits large frequency fluctuations. It also has the effect of reducing the overall size of the flywheel.

The fixed flywheel design allows for parallel operation of the units, with installation capacities of up to 5 MVA being a feasible proposition.

Depending on system requirements, the units can be equipped with manual and automatic no-break by-pass circuits for transfer of load to the mains in the event of maintenance or in response to possible faults.

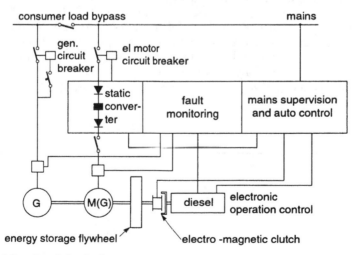

Figure 4.2 *Fixed flywheel system with static inverter*

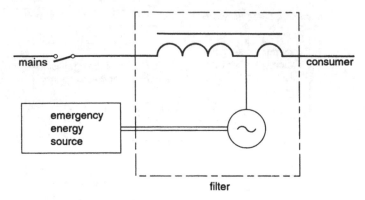

Figure 4.3   *Block diagram of Holec parallel system*

4.3 Induction coupling design

This design, known as a parallel system, takes into account the inherent reliability of the public mains supply. Connected in parallel to the mains, an emergency energy source (diesel engine) provides long term back-up in the event of power interruptions. A very effective but uncomplicated blocking filter is incorporated between the public mains and the load so as to prevent power line disturbances reaching the load. Thus optimum use is made of the high stability of the mains frequency.

This system, designed by the Heemaf Company in the Netherlands, was awarded a patent because of the unique connection to the mains supply. Heemaf became part of the Holec organisation during the 1960s and this design is now known as the Holec system.

The filter does not act to improve changes in frequency; the frequency at the input is the same as that at the output. However, large interconnected power system networks do not normally exhibit significant mains frequency variations. If the mains frequency is outside the tolerances imposed by the load, the emergency energy source takes over the power supply to the load without any interruptions. The filter stabilises the mains voltage in amplitude and phase symmetry. The filter consists of a series choke coil and AC compound synchronous generator; see Fig. 4.3 for a block diagram of the unit.

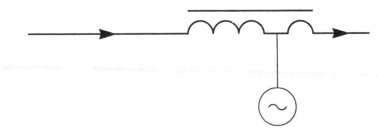

Figure 4.4   *Stabilising filter*

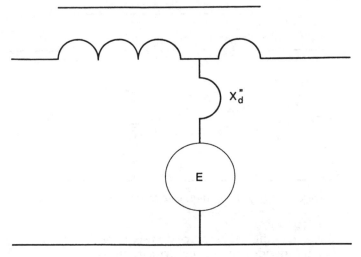

Figure 4.5   *Voltage source*

4.3.1 *Improvement of the quality of output power*

One of the major drawbacks of parallel connected systems in the past has been the performance under conditions of a short circuit at the mains terminals. Under such conditions voltage collapse could occur at the generator terminals. The blocking filter overcomes this drawback. It uses a coil which is tapped along its length (see Fig. 4.4).

The synchronous machine can be considered as a voltage source having an internal impedance which, at fundamental frequency, is equal to the sub-transient reactance X_d'' (Fig. 4.5). The relationship between the primary and secondary turns of the tapped choke coil is in the ratio 3:1. The reactance of the secondary is equal to the subtransient reactance X_d''.

In case of a full three-phase short circuit on the input terminals, voltage sharing will occur. In the sub-transient time range this sharing is shown in Fig.

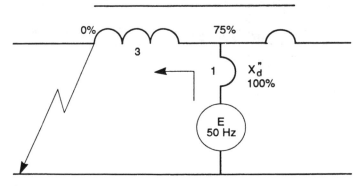

Figure 4.6   *Voltage sharing*

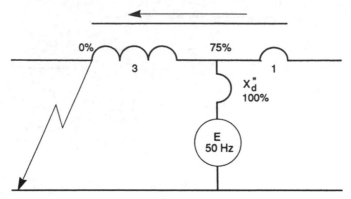

Figure 4.7   *Voltage at output terminals due to auto-transformer action*

4.6. The voltage sharing takes place between the input terminals and the voltage source and divides across the primary turns of the choke coil and the reactance of the machine.

At the tapping point the voltage is 75% of the nominal value. The autotransformer action of the choke coil results in a voltage of 100% at the output terminals (Fig. 4.7). Consequently the voltage at the output terminals remains constant.

The effect of the transient and very slow phenomena of the fundamental frequency on the consumer voltage are compensated for by the compounding of the synchronous machine, and can be totally eliminated by means of fine voltage control.

4.3.1.1   *Non-fundamental frequency*
For voltages which are not of the fundamental frequency, the synchronous machine will behave merely as a reactance. This reactance, X_2, is chosen to be equal to the sub-transient reactance $X_2 = X''_{dd}$ (Fig. 4.8).

Voltage variations are represented by a voltage source at the input terminals. The voltage at the tapping point is 25% of the input voltage at that frequency.

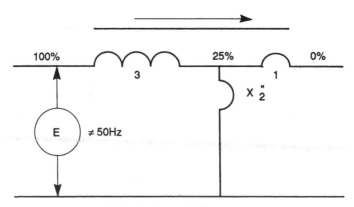

Figure 4.8   *Voltages, not of the fundamental frequency*

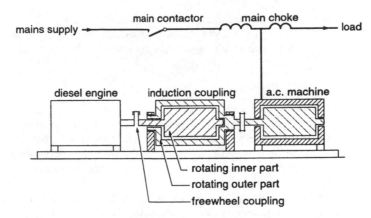

Figure 4.9 *Construction of a diesel UPS system*

By virtue of the auto-transformer action, as explained previously, there is a voltage drop from the tapping point to the load of approximately 25%. Thus the voltage is reduced to zero at the output terminals. In practical terms the deviation at the load terminals is well within the tolerances of the voltage sensitive equipment. Higher frequency signals injected into the mains, sometimes by the supply authority, are also attenuated such as not to interfere with the load.

The simple filter offers advantages in high efficiency during mains availability and a high reliability due to the minimal number of components through which the load current will flow. A further advantage is created by over-exciting the synchronous machine to produce reactive current. This causes an improvement of the system power factor to almost unity at the mains input terminals.

4.3.2 System configuration

The induction coupling type of diesel UPS consists of three basic elements, viz:

Filter unit: as described in Section 4.3.1
Emergency energy unit: A diesel engine which takes over supply of energy during mains irregularities and outages. The diesel engine is separated from the remainder of the UPS by means of a free-wheel clutch.
Induction coupling: The induction coupling is the device that stores energy sufficient to maintain power to the load whilst the diesel engine is started and brought up to speed. The induction coupling consists of two rotating elements:

(a) *Outer part*: Constructed as the stator of an asynchronous machines; mounted on a set of bearings and provided through slip rings with an AC winding as for an induction motor and a DC winding as for a DC motor. This outer part is driven by the synchronous machine (Fig. 4.9).
(b) *Inner rotor*: The rotor is a solid steel mass, dynamically balanced for 3000 rev/min operation. With AC supplied to the outer part, the rotor acts just as the rotor of an induction motor. With DC applied to the outer part, the

magnetic pull of the DC winding tends to lock the rotor to the outer part.

The inner rotor is connected via the free-wheel clutch to the diesel engine. Both the outer part and the inner rotor run concentrically inside one housing (Fig. 4.9).

4.3.3 *Principle of operation*

4.3.3.1   *Mains available* (see Fig. 4.10)

With mains available and within the tolerances of ±1% on frequency and ±10% on voltage, the filter system will maintain the output power from the UPS terminals according to the tolerances (Table 4.1). In addition the synchronous machine can supply reactive current in order to improve the power factor at the input terminals.

The outer part of the induction coupling is being driven at synchronous speed by the synchronous machine. The AC winding of the outer part, which is wound with 4 poles for a 50 Hz system, causes the inner rotor to rotate at nearly twice synchronous speed, i.e. 3000 rev/min. This is the source of kinetic energy for the time that it takes to start the diesel engine. By applying a DC field to the outer part of the induction coupling, the energy from the rotor is transferred to maintain speed of the outer part. The strength of the DC field is under control of the frequency regulator of the system. In this way a feedback loop controls the deceleration of the rotor.

Unlike the fixed flywheel system where only approximately 2% of kinetic energy is available, the induction coupling can make available 50% of its stored energy. Accordingly dimensions of the unit are relatively smaller.

4.3.3.2   *Mains irregularities or interruption*

If the mains supply exhibits irregularities such as interruptions, short circuits, too high or too low voltage, unequal phase voltages or frequency variations, then the system detects the irregularities and opens the incoming main contactor. The AC supply to the outer part of the induction coupling is interrupted.

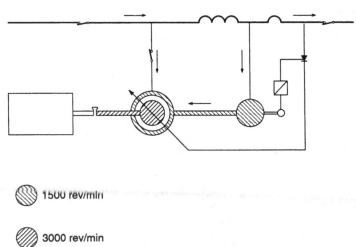

Figure 4.10   *Behaviour under mains conditions*

Table 4.1 *Technical data of induction coupling system*

Voltage and frequency characteristics
Static (steady state)
- At mains-voltage fluctuations up to ±10%
- At load surges up to ±10%
- At ambient temperature variations of 15 . . . 40°C

Maximum deviation from nominal value of
- Voltage: ±1%
- Frequency
  At normal operation: as with mains-voltage frequency
  At emergency operation: ±0·5%

Dynamic (transient)
- at mains failures
- at short-circuit in supplying mains
- at load surges of ±10 . . . ±50%

Maximum deviation from nominal value of
- Voltage
  during approx. 50 ms: −8% to +6%
  after maximum of 1 s: back to static value
- Frequency: ±1%

Voltage system symmetry
Maximum deviation of phase voltage and phase angle: 2% relative to symmetrical system, provided load asymmetry complies with

$$\frac{I_{max} - I_{min}}{I_{nom}} \leqq 20\% \text{ and } I_{max} \leqq I_{nom}$$

Waveform
Maximum harmonic distortion of generator voltage at linear loading: 5%

Filter operation
Reduction of higher harmonics from supply mains: 95%

Degree of protection
- Induction coupling: IP20
- Three-phase AC machine: IP22
- Equipment cabinets: IP21

Environmental conditions
Maximum operating altitude: 150 m above sea level
Maximum ambient temperature: 30°C
Maximum relative humidity: 70%

The diesel engine is commanded to start, and during its run-up time the DC winding on the outer part of the induction coupling is energised. The inner rotor is decelerated by the interaction of the DC field and the field associated with the

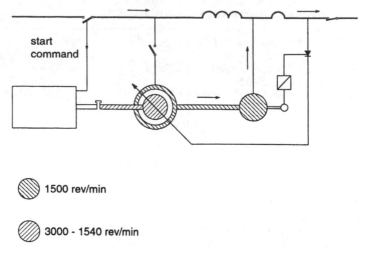

🌑 1500 rev/min

🌓 3000 - 1540 rev/min

Figure 4.11   *Behaviour under dynamic conditions*

eddy currents circulating within the inner rotor. Any tendency of the synchronous alternator speed to decrease is sensed by the frequency regulator. The regulator controls the firing angle of a thyristor bridge which controls the DC current flowing in the outer part of the induction coupling (see Fig. 4.11).

The energy transferred from the inner rotor to the outer part is equal to the load at that instant. The synchronous alternator speed is adjusted to meet the frequency requirements by means of the DC current regulation. During this period the diesel engine is at its no load speed of 1600 rev/min. When the inner rotor has slowed to 1600 rev/min the freewheel clutch engages, coupling the diesel engine to the inner rotor.

In simple terms; at the instant of mains failure the alternator is rotating at 1500 rev/min, i.e. a 50 Hz output.

The speed of the alternator will decrease; the rate of decrease being dependent upon the size of the load. The DC injected into the outer part of the induction coupling will have associated with it a magnetic field. In turn the eddy currents induced in the inner rotor will also have their own field. It is the interaction of the two fields which causes the induction coupling to act as a brake; in fact an eddy current brake. The braking action on the inner rotor sets up an opposite accelerating reaction on the outer rotor of the coupling.

The output frequency, i.e. the alternator speed, is controlled by the feedback loop which comprises a frequency sensing device to change firing angle of the thyristor. The DC current flowing in the induction coupling is therefore varied to adjust the braking action and hence the accelerating reaction.

4.3.3.3   *Emergency conditions*
The diesel engine drives the inner rotor at a speed of approximately 1540 rev/min. A DC supply to the outer part of the coupling transfers the energy, by means of the braking action, to the outer part and hence to the synchronous alternator (see Fig. 4.12).

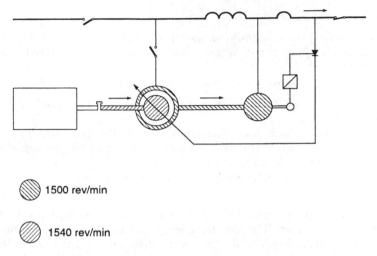

1500 rev/min

1540 rev/min

Figure 4.12 *Behaviour under emergency conditions*

The outer part frequency is independent of the speed variations of the diesel engine, provided the engine runs above a minimum speed. The control circuit adjusts the DC supplied to the induction coupling and maintains the output frequency within tolerance. In this condition the UPS system can run as an autonomous unit; assuming the role of standby emergency generator, albeit with sophisticated frequency control.

4.3.3.4 *Mains restoration*
The UPS system generates its own power during mains interruption, and therefore when the mains is restored there is likely to be a situation where the two supplies are not in phase. A paralleling check device senses this condition and waits until the vector systems are coincident before commanding the mains contactor to close.

The outer part of the induction coupling is energised with an AC supply and the inner rotor accelerates. At a speed of 1600 rev/min the free-wheel clutch disengages and the inner rotor continues to accelerate up to full speed. At full speed the rotor contains sufficient stored energy to deal with the next mains irregularity. After a period of approximately 3 minutes the diesel engine is switched off. The full operational sequence is shown in Fig. 4.13.

4.3.3.5 *Parallel operation and by-pass*
Each diesel unit can be operated in full parallel connection with other sets of the same rating to obtain redundancy and/or to increase output power (see Fig. 4.14). The synchronous machines are permanently connected to the supply and thus the principle of paralleling machines by a synchronising device does not apply in this case.

A fully automatic by-pass circuit is incorporated to increase reliability and to allow maintenance on single-set installations.

4.3.4 'No break KS' system

Evolving from the previous design is the 'No-break KS' system. This is based on similar principles but the two electrical machines are replaced with one combined unit.

The electrical machine, known as a stato-alternator consists of:

- An outer casing which includes a conventional armature winding, and two exciter windings
- A main rotor which has associated with it the pole windings of the synchronous machine, the armature of the two exciters, an asynchronous brake, rotating rectifiers and a cooling fan
- An internal rotor which incorporates the squirrel cage rotor of the brake, and a flywheel.

When mains failure occurs the outer part continues to feed the load as a synchronous alternator. The kinetic energy and braking energy continue to drive the alternator. Meanwhile the diesel engine is started electrically. The electromagnetic clutch engages approximately one second later and the diesel completes its acceleration. After a suitable time delay, approximately ten seconds, the inner rotor of the machine is accelerated back to its full operational speed.

In some cases it is feasible for the KS system to also supply an element of short-break power in addition to its no-break power. By suitable oversizing of the engine and clutch, a supplementary load, not in excess of the no-break load, can be supported after a break of approximately ten seconds. In this condition the unit acts as a no-break supply and a conventional stand-by generator.

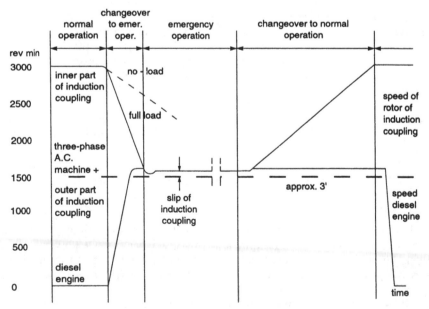

Figure 4.13   *Time sequence diagram*

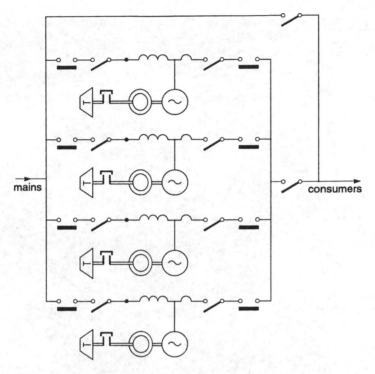

Figure 4.14 *Connection for parallel redundant operation*

4.4 Reliability considerations

The important factor when considering a load which is deemed a critical load is the continuous operation of the load; i.e. for a computer installation it is Uninterrupted Computer Operation not Uninterruptible Power Supply which is the important issue. The only true guarantee that a manufacturer can give is: 'we guarantee that this machine will fail'. The task of a system designer is to evaluate possible areas of failure and to minimise the effects on the system as a whole.

At the design stage of a UPS installation it is therefore necessary to consider load requirements; future load requirements; provision of redundant plant to account for failures in individual modules. By means of theoretical reliability calculations relative to the number and nature of consistuent elements it is possible to arrive at statistical information relating to mean time between failures (MTBF).

Diesel UPS systems offer advantages in the form of their ruggedness, ability to provide a sinusoidal output, and space-saving opportunities due to there being no need for large battery banks. Consequently for the larger UPS installation the diesel systems are worthy of serious consideration. In simple terms the diesel UPS can be considered as an emergency stand-by generator with means of providing power conditioning and true no-break power supply.

Figure 4.15   *Holec Galileo UPS installation. ©1988 Old Town Studio, Swindon*

The parallel connected system, as detailed in this chapter, has an inherently low source impedance. This is an important feature when supplying non-linear loads (see Chapter 10).

A series connected system, where the UPS is effectively in series between the mains supply and the load, can result in a system which is less reliable than the mains it is supposed to protect. A parallel connected system, such as the induction coupling design, uses the inherent reliability of the mains supply and hence maximises the reliability of the installation.

A further factor which relates to both types of diesel UPS is the high short-circuit capacity of the alternator. This ensures that down-stream fuses are cleared even during emergency operation without degradation of the output of the UPS. The subject of reliability is dealt with comprehensively in Chapter 11.

In practical terms we are dealing with a system which relies on rotating machinery. Obviously maintenance is a necessary requirement, but there is the advantage of actually being able to carry out preventive maintenance.

A crucial factor is the starting performance of the diesel engine. The start operation is carried out using an electric starting motor fed from a heavy duty battery. The status of the battery is constantly monitored and an alarm is given should the battery voltage be impaired. An electric immersion heater in the water cooling jacket of the diesel engine maintains the coolant at a temperature of 40°C. Thus the engine is primed, ready for use.

In the induction coupling (Holec) system, the free-wheel clutch effectively separates the diesel engine from the high inertia items, such as the alternator. Thus when the engine is commanded to start there are no high inertia components connected to the engine. This situation allows for the rapid start time of the diesel engine (approximately one second). In field practice we see that the diesel engine exhibits a failure to start statistic in the order of 1 in 6000. This figure is for all starts, including those made for maintenance purposes.

Therefore in practical terms the reliability of the diesel engine system relates to a comparison between the availability of a large number of cells in a battery bank and the start performance of a monitored diesel engine. The inherent reliability of rotating plant, the ability to provide fault current from within its own resources, and the opportunity to carry out preventative maintenance combine to present the diesel UPS as a valid solution to critical power problems.

4.5 References

BECK, B. J., and LAYCOCK, S. M.: 'Operation and reliability of diesel no-break UPS systems'. Power 87 Conference
BECK, B. J.: 'Uninterruptible power supplies and the Heemaf system', *Electron. & Power* July/August 1981

Chapter 5

Static thyristor inverter system with battery energy store

M. J. LEACH

5.1 Introduction

The UPS of 2 MVA is not uncommon as a power source for modern large computer suites. Banks, building societies and credit card companies are all examples of large computer users whose profitability can be disproportionately affected by a short interruption in supply. Often the computers are linked by modem to other users, so multiplying the effect. The UPS also isolates the computer from the noise on the supply; noise which can cause data errors and unexplained hardware failures. With such a large power requirement, there are only a small number of switching devices that can be used economically. But one must look at other aspects of the UPS before making a choice.

The large UPS is loaded with many small loads, mostly using switched mode power supplies (SMPS), drawing their power over a small portion of each half cycle. These single-phase loads are grouped together into the three-phase load for the UPS. The harmonic content of the current waveform is rich in odd harmonics; predominantly third, fifth and seventh. This is true also for the smaller UPS but the solution to the problem of how to supply such harmonic currents is different. At low powers, pulse width modulation (PWM) with subcycle distortion correction is used to reduce the resultant voltage waveform distortion. To do this at large power levels would require a switch with high operating frequency, high voltage and current capability, low switching losses and low storage times. Of course, many transistors may be operated in parallel or whole switching units paralleled but the cost or complexity do not justify this approach. At the present time the economic break point is about 300 kVA and this is only justifiable when the UPS must be located in the computer room.

Another solution to the problem of the load current harmonics is to use a filter with low impedance at the main harmonic frequencies. This can take the form of traps tuned to say the fifth and seventh harmonics (third and multiples can be given a zero impedance path in a three-phase connection).

With PWM switching, the filter size can be reduced to a minimum, but as power requirements rise the devices have to be used at lower repetition frequencies and the advantage is lost. A good cost effective solution is to use a switch operating at the fundamental frequency coupled with filters for the major harmonics in the resultant square wave. These same traps will provide the low impedance required by the load.

With this strategy, the choice of switch becomes easy. Only the thyristor combines high-voltage high-current capability with low cost and low turn-on power. The disadvantage is that it is difficult to turn off. It is more than ten years since the first gate turn off (GTO) thyristor was used in a UPS but the cost of the device is uneconomic except for the few UPS manufacturers who also manufacture semiconductors. It is interesting to note that some manufacturers of large variable speed drives who adopted the GTO with are now rethinking that decision.

5.2 Problem of thyristor turn off

As a switch for switching ON a current, the thyristor is almost ideal, having a gate current of less than 1 A from a source of less than 3 V switching a current of more than 1000 A from 1000 V! The problem is that it will not revert to its blocking state until the current has been reduced to zero for a short time. There are many ways of achieving this; mostly based on resonant circuits or auto transformer action [1]. The added components make the inverter larger but the overall design is a good compromise between cost, size and performance for larger UPS.

5.3 Rectifiers

Large UPS usually work on a DC link voltage of 300–500 V in order to optimise the constraints of transformer cost, high DC currents and inverter commutation constraints. At 450 V there is no need for an expensive transformer between the 380/415 V line and the rectifier. In UPS, reliability is one of the important goals the designer must achieve. Reducing the number of components is one way of achieving this, so the simple three-phase six-pulse bridge is attractive.

A diode bridge is fine when no control of the DC link is required. However, this implies that the battery must be electrically isolated from the DC link and be separately charged. A means must be provided to reconnect the battery on loss of supply. A controlled DC link with a battery floated directly across it trades the complexity of a separate charger and switch for the relatively simple phase controlled thyristor bridge. So a six-pulse phase controlled bridge has become a popular choice for large UPS.

In designing the phase controller circuits for the rectifier, it is necessary to recognise that small dips and outages will occur to the supply system and the controls must be tolerant to these. Also, the zero crossing reference points may be distorted by noise or commutation notches. Phase angle relationships may change as faults occur on other users' equipment connected to the same line. Modern control circuits using 'flywheel' principles or the ubiquitous microprocessor can be very tolerant of these effects.

Simple bridge rectifiers, whether controlled or not, draw harmonic currents from the supply and thereby cause voltage distortion. The amount of distortion depends on the point of common coupling. A UPS of 500 kVA with a dedicated supply transformer from 11 kV to 415 V can use a simple six-pulse phase controlled rectifier. The same unit could cause unacceptable supply distortion on a shared 415 V supply. The classical way to improve the situation is to use a 12-pulse rectifier. This draws no current at the fifth and seventh harmonic but requires an expensive double wound transformer for phase shifting and twice the number of SCRs. This technique is still used when a double wound transformer is required for other reasons but there are alternative methods.

5.4 Input filters

As shown in Chapter 10, a bridge rectifier would ideally take a square wave of current in each diode or thyristor with the resultant high percentage of odd harmonics. Also shown is the equally undesirable commutation notch. This can be a nuisance to other users of the supply. Other equipment using phase control can jitter in phase angle when a notch coincides with a sensing point or commutation point. It can add to the high frequency harmonics or cause interference to telephones and other audio equipment.

For these reasons it is desirable to use series inductors to reduce the higher harmonics and to offer optional filters to limit the distortion further. Current distortion of a six-pulse bridge is around 30% and filters can reduce this economically to 10–15%. A simple series inductor/shunt capacitor will achieve this but the capacitor may be rather large. This in turn gives a degree of power factor correction — useful at full load but could be excessively leading at light load.

A better way is to use trap filters tuned to the major harmonics. Placed between the supply and the rectifier, the trap offers low harmonic impedance but must work in conjunction with inductors placed on both the supply and rectifier sides. Without the line side inductor, the UPS input filter would soak up the harmonic currents generated by other users of the supply — a general clean-up device the supply authorities would surely welcome!

5.5 Alternative to an input filter

The UPS rectifier can work directly from the 380/415 V supply but a small auto-transformer is often used to give the correct voltage/power factor combination for the chosen DC link voltage. However, this transformer is only a fraction of the size and cost of a double wound transformer and the additional cost of a small amount of phase shifting is also low. When several UPS modules are connected to the same supply, the different phase shifts on each module can be arranged to cancel or reduce the majority of harmonics. With many modules, only a small amount of phase shifting is required in each. Careful design can ensure that only a small increase in input harmonics occurs when any one of the rectifiers is switched off. The effect is shown in Fig. 5.1. With one UPS rectifier, the line current consists of two pulses contained with the theoretical square wave. As more modules are added, the total current becomes more sinusoidal.

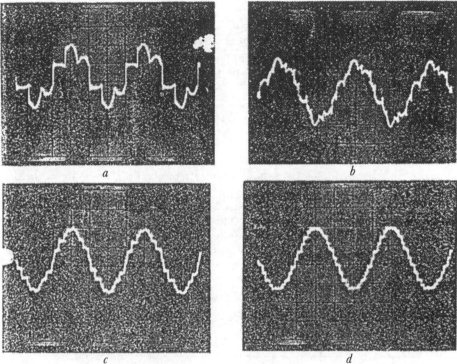

a   *b*

c   *d*

Figure 5.1 *Input supply current waveforms showing the effects of phase shifting for harmonic cancellation*

(*a*) One, (*b*) two, (*c*) three, (*d*) four UPS modules

5.6 Battery charging

It is usual to float charge the battery directly across the DC link between the rectifier and the inverter. Correct choice of float voltage is important for the type of battery chosen. For lead acid this is around 2·25 V per cell for wet cells and 2·275 V for sealed cells. With nickel–cadmium batteries it may be necessary to use a slightly larger battery than optimum at a lower voltage per cell to avoid excessive water consumption.

In the float regime, the phase controlled rectifier must also be controlled by the battery recharge current. Once the battery current limit has been reached, the link voltage will be kept low. It automatically rises as the battery recharges. Note that, once normal float voltage is reached, that is not the end of battery recharge. The battery will still need a considerable time to be fully recharged. A good guide is that 95% of the energy will be replaced in ten times the discharge time. There are many arrangements of batteries and rectifiers to suit all pockets and reliability targets. Each UPS may have its own battery or a single battery may supply many UPS modules (see Fig. 5.2).

In very large systems it will often be necessary to place batteries in parallel to achieve the required capacity. It then makes sense to take the individual batteries and use one for each of the UPS modules. Redundancy is then improved and maintenance of the battery (for example, topping up with water) does not remove all the battery from the system. In systems made up from a number of small UPS modules in parallel, a common battery feeding (and being recharged from) all the modules can offer the cost savings of a large capacity cell.

5.7 Voltage control

The need is to control the UPS output voltage against regulation drop and the variable DC bus voltage in a float charge regime. In UPS systems below 300 kVA, it is possible to isolate these two functions. Particularly in small UPS, a separate DC–DC convertor can be used to maintain constant the inverter bus during battery discharge. However, efficiency is reduced by converting the power twice and this is particularly undesirable in large systems.

With a square wave inverter, voltage control can be effected by having not one but two inverters. They are phase shifted relative to each other and their

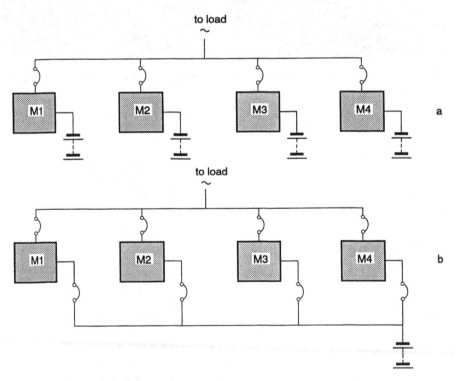

Figure 5.2   *Battery configurations*

    a Individual batteries
    b Common battery

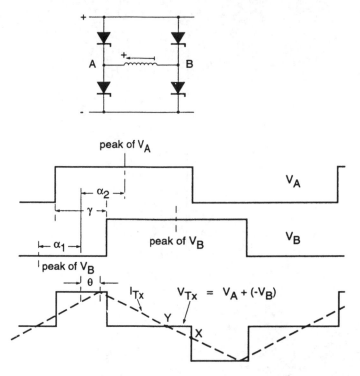

Figure 5.3 *Switching pattern for two phase shifted inverters used for voltage control*

output is summed. Control covers the whole range from zero to maximum voltage. This large range is useful during start up to 'walk-in' the AC output voltage, or during fault conditions to reduce the output quickly. Note that it is necessary to phase shift both inverters an equal amount in opposite directions relative to the synchronising or reference waveform. In the short time the bypass supply and UPS are in parallel, a massive circulating current would result if this was not done.

With phase shifted square wave inverters, the power is shared between the two inverters but it is not divided equally. Fig. 5.3 shows the arrangement of the two inverters for one phase. To produce zero voltage output, the two inverters are switched in the same pattern whilst to give maximum voltage they are switched in anti-phase. At any point between, the thyristors of inverter B will appear to be following the pattern of thyristors A after a delay period. The waveforms are shown below. The vector diagram (Fig. 5.4) shows how the leading inverter A has a leading current whilst the lagging inverter B (inverted to $-B$) sees a considerably lagging current ($\alpha 1 + \Theta$). Advantage can be taken of this fact to reduce the commutation ability of the leading inverter. A leading current implies that the current has decreased to zero or reversed before the commutation point is reached.

There is, however, one disadvantage of phase shifted inverters. The ratio of the odd harmonics to the fundamental changes with the phase angle or output

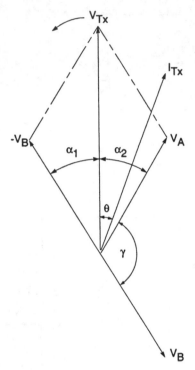

Figure 5.4　*Vector diagram for the system shown in Fig. 5.3*

voltage. The effect is easily calculated and is shown graphically in Figs. 8.3 and 8.4 of Reference 1. Of course, at certain angles the harmonics cancel to zero but the design must allow for the worst case.

5.8 Output filters

As mentioned earlier, the purpose of the filter is twofold. Firstly it must reduce the harmonic content of the UPS output voltage to an acceptable level and secondly it must have a low impedance for load generated harmonic currents.

Traps set at the two lowest harmonics can reduce the total harmonic distortion (THD) to less than 5% if coupled with an attenuating impedance for the higher order harmonics. In smaller SCR UPS, one manufacturer has made use of a broad tuned trap set at the sixth harmonic as a means of reducing cost and component count.

In the past, UPS used filters tuned to the fundamental. A series tuned inductor and capacitor were placed between the inverter and the load and this was supplemented by a shunt arrangement of a capacitor and inductor, again tuned to the fundamental. The resultant waveform is very nearly sinusoidal with good THD. Unfortunately, the component values need to be large and the energy storage is high. With dynamic loads on the UPS, such as motors, energy transfer can take place in a cyclic manner between the filter and the motor. However, lower cost is the main reason that alternative filters are used today.

5.9 Static bypass

The static bypass is a means of quickly bypassing the main parts of the UPS by connecting the mains input to the load terminals. It consists of anti-parallel SCRs in each phase and a (usually independent) power supply and firing circuit.

A static UPS can never have the energy storage ability of a rotating system to clear sub-circuit load fuses. However, the reliability of the mains is more than good enough to be considered as a power source for this task. The chance of a load fault and mains failure at the same time is very low. To clear the load sub-circuit fuses, the static bypass can be fired for a cycle or more without opening the breaker to the inverter. Effectively, the mains and the UPS supply are in parallel but this is permissible in a well designed system. This is one of the three uses of the static bypass.

The second use is to connect the load to the mains supply if the UPS should fail. Again, even if the mains is considered only to have a mean time between failures (MTBF) of 100 hours, this feature will add significantly to the total system reliability.

The third use is as a bridge across the main circuit breakers during normal change over. On starting the UPS or during maintenance, the loads will be connected to the supply through a bypass circuit breaker. The static bypass switch can be fired to, in effect, convert the two breakers to 'make before break' (see Fig. 5.5).

5.10 Alarms and interfaces

Large UPS are usually placed in the basement or a closed equipment room. Originally, this was a necessity as the large batteries had to be located near the UPS and lead–acid batteries needed special rooms and equipment. With modern sealed re-combination batteries there is not the same requirement. The batteries are often placed in corridors or convenient places now that electrical safety is the only criteria. This has allowed the UPS to be squeezed into small spaces too. However, there is still a need for remote alarm and indication of UPS status. Full 'on-board' metering and mimic diagram are not as popular now that UPS are considered reliable products.

Most requirements are met by alarm contacts in the UPS module. The main need is for the data processing manager to know that the UPS is 'healthy', whether running on mains or battery, and how long he has left when on battery. In a large installation there will be a maintenance department or a security department that will require remote alarm indication of faults or irregularities.

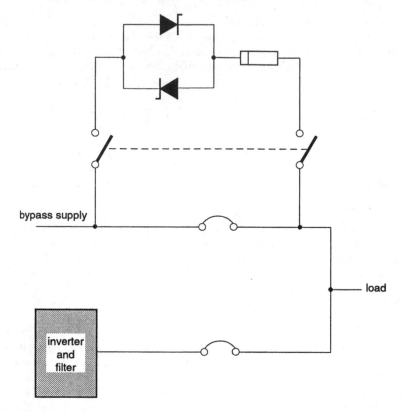

Figure 5.5   *Static bypass arrangement*

The following is a typical list of alarms:
- Mains failure
- Battery circuit breaker open
- DC overvoltage
- Overload
- Overload shutdown (for protection of the UPS but the load is transferred to the bypass)
- Emergency shutdown (from slam switches in the computer room and plant room or coupled to the fire control system)
- Fan failure (loss of a fan is not catastrophic as these are usually in a redundant configuration)
- Output voltage error
- Control power failure (usually two power supplies — one fed from the mains and one from the UPS output)
- Battery on load
- Low battery
- Equipment over temperature
- Load on bypass
- Static bypass inhibited (because the supply is outside the UPS output frequency specification)

In addition to these there is always a 'common alarm' which comes on when any alarm is present, and a cancellable 'new alarm' which is activated whenever a second or subsequent alarm is initiated.

The trend is to reduce the need for maintenance and reduce manning levels at sites. This increases the need for more indications that can not only sound a bell or flash a light locally but can be sent down a communications link to a computer terminal or a telephone line. In very large systems, local alarms are sufficient. It is in mid-sized systems where RS232 serial data transmission is becoming commonplace. Once a data link has been established, it can be used for remote diagnostics too. The 'clever' UPS is just beginning to appear. Like the computer that automatically telephones its manufacturer when it discovers a software error, the UPS in future will telephone the manufacturer's service department to request a visit or even diagnose the fault and order the correct spare parts.

5.11 Multi-module systems

To increase the available power or to give a degree of redundancy, UPS modules are frequently connected in parallel (see Fig. 5.6). It is imperative that the modules not only run at the same frequency, but are in phase with each other and have the same output voltage. They must also share the load current equally between them.

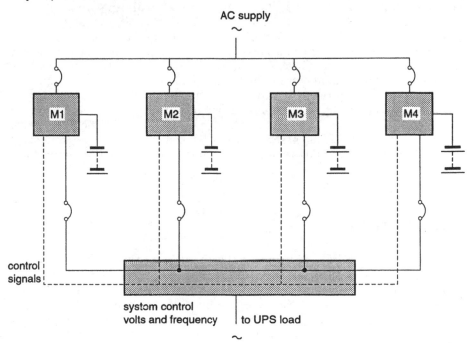

Figure 5.6   *Connection arrangement for a multi-module system*

To the designer, this a formidable task if he is not to introduce common points of failure that will reduce the overall reliability. Signals must pass between the UPS modules in such a way that loss of any one signal does not jeopardise the system integrity. More important is that a false signal from a faulty module does not 'fool' the whole UPS system into operation out of specification. An example of this might be in synchronisation. A faulty module could send out synchronisation pulses at a high frequency and cause the other modules to follow.

Large fault currents can build quickly between good and faulty parallel module. The fault must be detected and isolation achieved within a few hundred microseconds if the output voltage is not to be unduly disturbed. Special high speed breakers with large coil overdrive have been used but a static semiconductor switch is faster. One popular UPS range uses energy stored in a capacitor to clear a series fuse. A resonant circuit, which includes the capacitor, is switched on by a large SCR and the resultant high current pulse soon opens the fuse. This is done simultaneously on at least two phases.

One UPS manufacturer has not only a fail-safe signalling system but offers an optional voting system at the system level. In this arrangement, each of the paralleled modules compares its own performance with those of its adjacent neighbours and sends the results back to the system cabinet. Voting logic then decides when a UPS module is faulty and quickly isolates it before fault currents can grow. For example, if module 3 thinks it is performing the same as module 2 but not the same as module 4, whilst module 5 thinks it is performing the same as module 6 but not module 4, then the voting logic will trip out module 4 no matter what it thinks.

Such systems also have to be designed very carefully to avoid jeopardising overall reliability. They must also be tolerant of modules being taken in and out of the system for service. Often this option is chosen in 400 Hz systems where there is no backup available from the bypass. There is often a system cabinet to act as the marshalling cabinet to connect the modules in parallel. This is a good place to put the system metering and a system static bypass. In multi-module systems it is usual, but not always so, that the static bypass is removed from the individual modules. It is here too that the control signals are connected together, the system to mains synchronisation is performed and the voting logic is placed.

5.12 The future

Large UPS modules have come a long way in the last 20 years. One has only to look at the volume occupied by, say, a 100 kVA unit to see the immense changes that have occurred in design (see Fig. 5.7). A single module 800 kVA UPS with static bypass now occupies a volume of only 8 m^3 (100 kVA/m^3).

In the same time, reliability has improved to a point where many computer users do not realise they have the protection of a UPS. Maintenance and fault diagnosis have changed from being a regular job performed by the customer's own manufacturer-trained maintenance department to being an annual health check by the manufacturer's service department.

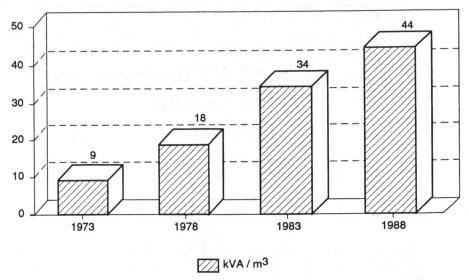

Figure 5.7 *Improvements in 100 kVA UPS design over 15 years (in kVA/m³)*

The trends continue. Large UPS will not grow in popularity at the same rate as their smaller cousins but the reduced costs open up new application areas. Already, large UPS are being placed at the supply input of factories to give total protection against loss of supply to the machinery. This often occurs in a Third World country where supplies are inconsistent and modern Western machinery is in use.

Improvements in semiconductors may allow faster switching frequencies in future without the penalty of high losses. Then we may see PWM techniques adopted at high powers. The computer industry is now coming to terms with the harmonic pollution caused by switched mode power supplies commonly used in all sizes of computer. Their change over to power supplies drawing a sine wave of current may be mirrored in a requirement for UPS to be similarly kind to the supply system. Communications will improve to the main frame computer being served and to remote diagnostics and data collection points.

In fact, the large UPS is becoming a mature product and the future should entail a steady improvement in performance and a reduction in size and cost.

5.13 Acknowledgments

The author wishes to thank Emerson Electric Co., Computer Power Division for the provision of documents and information.

5.14 Reference

1 BEDFORD and HOFT: 'Principles of inverter circuits' (John Wiley & Sons, Inc. ISBN 0 471 06134 4)

Chapter 6

Static transistor UPS incorporating battery backup

E. P. BARNETT

6.1 Introduction

The ever increasing use of computer systems, especially in the medium power range (10 MIPS, 5 Gbytes), results in an ever increasing requirement of associated UPS systems generally with ratings of up to 100 kVA. In many of these applications the reliance placed on the system dictates that a secure clean continuous power supply is available.

They are often installed in existing buildings in which there is no dedicated plant room. The alternative is to locate these systems within the computer room. Obviously the traditional large system, producing in excess of 70 dBA noise level with high heat losses, is not suitable for operation in this type of environment.

There are many aspects of the design of these systems which differ from the plant room systems. Obviously the first criterion with any system must be reliability, the noise emitted must be less than the noise level in the computer room, and the heat losses reduced by designing a high efficiency system. With the module located in the vicinity of the computer equipment much more consideration must be given to radio frequency interference (RFI) either conducted or radiated, both of which can result in data corruption.

Whilst the system will be three-phase input and output, there will normally be a large content of single-phase load, and hence it must be able to withstand the possibility of unbalanced loads; it will inevitably have some non-linear content.

If, in the design of the system, consideration can be given to diagnostic facilities and component access this can reduce the repair time in the event of a failure. The availability of improved communication techniques now allows fault finding to be carried out remotely, again resulting in a reduction in down time.

The manufacturer has to try and achieve all these features whilst still offering a competitively priced product. The system should comprise a true on-line system (Fig. 6.1) in which the power is continuously conditioned. The additional benefit of this is the security of knowing that the unit can supply the power, rather than using an off-line, or bypass priority system, in which the unit supplies only a portion of the power under normal operation.

Under normal operating conditions the power supplied from the mains is

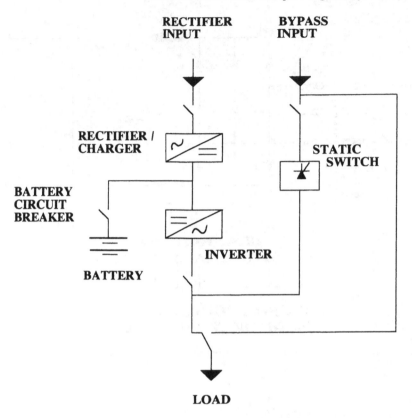

RECTIFIER INPUT   BYPASS INPUT

RECTIFIER / CHARGER

BATTERY CIRCUIT BREAKER

BATTERY

STATIC SWITCH

INVERTER

LOAD

Figure 6.1 *Single line UPS system*

rectified and fed to both the inverter and the battery. The inverter produces from the DC a true sinusoidal waveform, and, in the event of breaks in the mains supply, the energy stored in the battery is utilised to provide the inverter input power requirements. In this way deviations, breaks and distortion of the input supply are totally eradicated. This is a true on-line system of the type which is recommended and preferred by the major computer manufacturers.

The static switch provides an alternative source of supply for the load in the event of equipment failure or short term overloads. The manual bypass switch enables the system to be totally bypassed for maintenance purposes.

It is not only the development of semiconductor power devices but also the improvement and acceptance of sealed lead–acid batteries which has enabled us to locate UPS systems in the computer room. The battery is a vital component of the system, and within the design parameters consideration must be given to its charging and to the environment in which it is housed.

However, let us consider in greater detail the individual sections of the system.

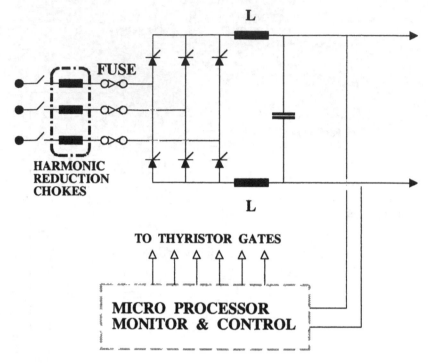

Figure 6.2 *6-pulse charger*

6.2 Charger/rectifier

This consists of a three-phase fully controlled thyristor rectifier bridge (Fig. 6.2) in which the conduction point of the thyristors is varied to control the DC output voltage (Fig. 6.3).

Information relating to the charging voltage and current, monitored across and through the series shunt, is fed into the microprocessor, which in turn controls the conduction angle of the thyristors. In this way we are able to control the charging voltage to within 1%, a most important parameter when the system is used in conjunction with the sealed lead–acid battery, as an increase in voltage may well give rise to overcharging and subsequently damage to the battery, whilst a reduction will result in undercharging and a loss of capacity. The charging current is also controlled very accurately; it is normally set to 10% of the 10 hour capacity. Whilst batteries are able to accept greater charging currents than this, very little advantage is gained by recharging at the higher levels. Fig. 6.4 gives an indication of capacity replaced at these differing rates.

The mode of charging is known as constant voltage current limited (Fig. 6.5). Following a discharge, initially the current is limited to the predetermined level above; this is achieved by reducing the charging voltage. As the cell voltage rises to the normal float charge level of 2·27 V per cell the voltage is then limited and the current decreases.

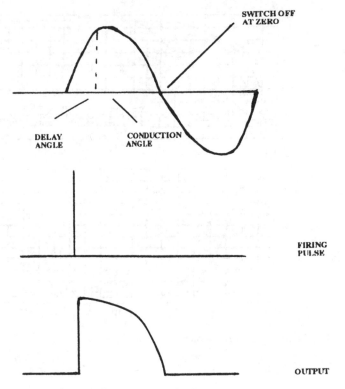

Figure 6.3   *Thyristor input/output*

The charger/rectifier must obviously be rated to supply the maximum input power requirements of the inverter, together with the power required to recharge the battery. The output regulation under all conditions of load must be good (Fig. 6.6). Poor regulation can result in a condition known as 'capacity walk down' whereby, during the time the inverter draws pulses of current from the rectifier, the output voltage falls; effectively voltage ripple is superimposed on the DC. The result is that the battery now supplies some of the inverter power requirements. This phenomenon is also seen when chargers have inadequate smoothing, and again AC voltage ripple content is superimposed on the DC voltage, whilst the discharge occurs for only a very short duration. Because the discharge efficiency is much greater than the recharge efficiency, especially when the battery is 90–100% charged, the overall effect is a slow reduction in the available capacity.

The voltage ripple effectively causes current ripple so far as the battery is concerned; and we have cycles of discharge and charge. Obviously the degree of current ripple will depend on the voltage deviations and the load current. It will, however, give rise to heat generation within the cells. The battery manufacturers limit the ripple to 7% of the C3 (3 hour) capacity or 5% of the C10 rate. Under these conditions the temperature rise within the cell will be limited to a level acceptable to the manufacturers, which is normally a maximum of 5°C.

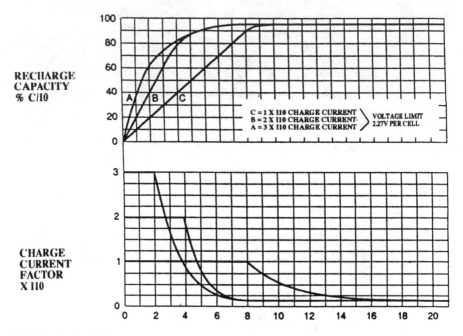

Figure 6.4 *Recharge at various rates*

The frequency of the ripple is also significant as far as the heating effect of the battery is concerned, the most damaging frequency being between 100 and 250 Hz. The ripple frequency from the charger will be 300 Hz and that due to the inverter will be a function of the switching speed of the inverter, which will typically be 1800–5000 pulses/s. Hence the possibility of damage from these sources is considerably reduced.

The type of charger used generally is six-pulse full wave with output filtering. It will always limit the voltage ripple to not more than 1% and the subsequent current ripple to within the limits stipulated above. The ripple superimposed owing to the inverter can also be negated by fitting a choke in the circuit between the battery and the inverter input (Fig. 6.7).

Pollution of the incoming supply due to the harmonic currents drawn by the charger is another facet of the design which needs to be given consideration. G5/3 issued by the Electricity Council gives recommendations regarding the degree of pollution which is allowable. Two methods are available for the reduction of these currents: one uses a 12-pulse rectifier system (Fig. 6.8), in which two six-pulse bridges are phase shifted from each other, resulting in the cancellation of the fifth and seventh (and multiples) harmonics. This solution reduces the RMS current distortion to approximately 12% of the fundamental.

The alternative is to fit a filter (Fig. 6.9), normally tuned to, or near, the major harmonic (fifth) but giving a reduction in all the harmonics, dependent on their proximity to the fifth. The overall reduction is to approximately 5–10% of the fundamental. Both methods offer a number of benefits; for example, the filter, fewer components, greater efficiency, greater reduction, and the ability retrospectively to fit the filter. They can if required also be designed to provide power factor correction, thus reducing the running costs. Care must be taken with lightly loaded systems to avoid the possibility of leading power factors.

The use of thyristors in the rectifier bridge also gives the possibility of 'ramping up' the input power (Fig. 6.10) over a period of 15–20 s following the return of the mains or generator supply. This is achieved by again adjusting the conduction angles of the thyristors. It is especially important when UPS systems are used in conjunction with generators, as high impact loads on generators can result in voltage and frequency fluctuations which may well result in the supply being out of tolerance as far the rectifier is concerned.

6.3 Batteries

Sealed lead–acid batteries are used in almost every application of UPS systems with ratings up to 100 kVA. It is only during fault conditions when the battery is overcharged that any gas is given off. Whilst the cells do not have a vent in the traditional sense, they do have a safety vent which releases should there be a build up of gas within the cell. Originally sealed cells were sold as totally maintenance free, but they do, in fact, require a visual inspection, cleaning, checking of the termination tightness on a regular basis and their cell voltages must be recorded.

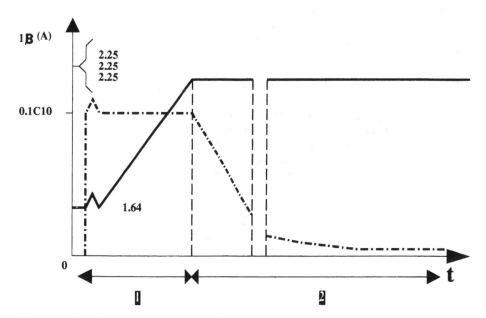

Figure 6.5 *Constant voltage/current limited charging*

1 current limited 2 constant voltage

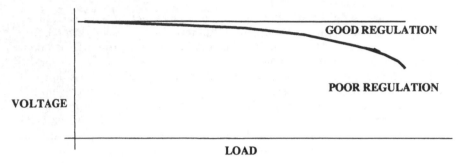

Figure 6.6   *Output regulation*

The quality and life expectancy of the cells varies from 3 to 10 years, depending mainly on the plate thickness, but is very much influenced by the environmental conditions in which they operate. Ideally they ought to operate within a temperature range of 20 to 25°C. In temperatures below 20°C the performance starts to fall off, whilst above 25°C the performance improves owing to the increase in the efficiency of the chemical reactions, but this is at the cost of life, as the corrosion rate also increases. As an indication, for every 5°C above 25°C the life expectancy will be halved. Hence the location of the battery within the air-conditioned environment of the computer room ought to enhance its life, given that an adequate charging regime is used and the battery cubicle is provided with adequate ventilation.

Obviously the capacity of the battery and the physical size are very much dependent on the standby time required and the power of the UPS system. In most instances a 10 min autonomy is selected with longer term standby provided by a diesel generator supply replacing the mains source.

6.4 Inverter

The basic switching action of inverter systems, be they transistor or thyristor, is similar. The bridge usually consists of four main power devices per phase, two of which are switched on at any instant (Fig. 6.11). On closing S1 and S4 current will flow through the load. Conversely, closing S2 and S3, current again flows through the load; however the polarity across the load is now reversed.

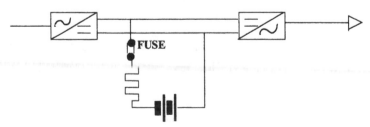

Figure 6.7   *Harmonic choke in battery circuit*

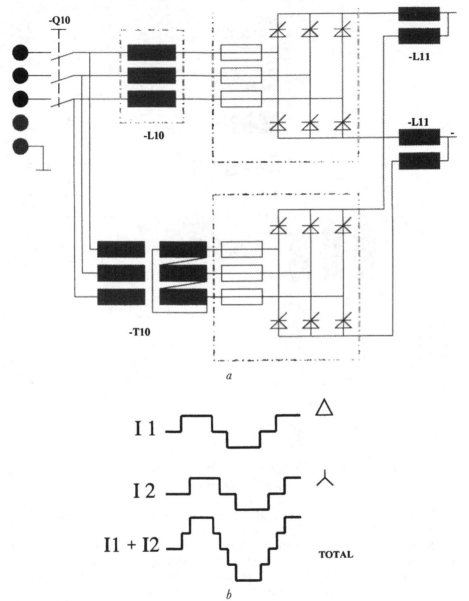

Figure 6.8 *12-pulse charger system*

(*a*) 12-pulse bridge
(*b*) Input current to 12-pulse bridge

In practice the switches are replaced by power transistors and the load by the primary of the output transformer (Fig. 6.12). The ease with which transistors can be switched on and off gives a considerable advantage over the thyristor inverter. The decision which has to be taken is at what frequency the switching

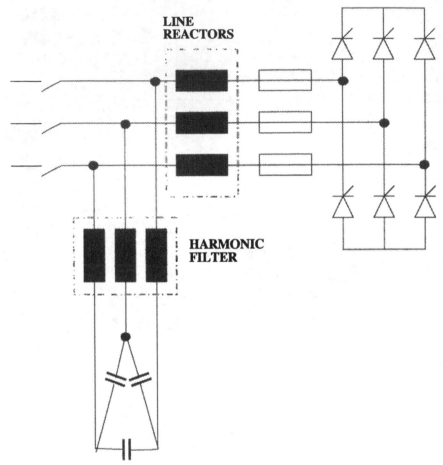

Figure 6.9   *Input harmonic filter*

or chopping of the DC occurs and the method of achieving it; but the scope is somewhat limited by the power devices available.

The importance of the switching speed is related directly to the noise level produced by the system, as most of this originates from the stressing of components during the switching pulses. If, for example, we could switch at above 15 kHz the frequency of the noise produced is above the hearing range of most of us. The difficulty arises, however, when we try and find a transistor which will successfully switch currents of sufficient magnitude at this frequency. Given a nominal DC voltage of about 400 V the corresponding current for systems with ratings of 10–100 kVA would be of the order of 30–300 A.

The power handling capability of transistors is directly related to their physical size, as is their ability to switch at high speeds. Hence, the choice we have, if we wish to use high frequencies, is to use a large number of low power devices connected in parallel, or connect a number of smaller power inverter stacks via coupling transformers. Both solutions, however, tend to reduce the

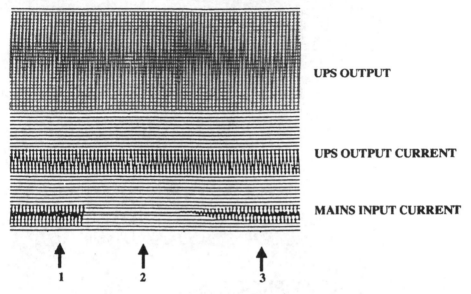

UPS OUTPUT

UPS OUTPUT CURRENT

MAINS INPUT CURRENT

1 2 3

Figure 6.10 *Ramp-up of input current by control of thyristor*

1 mains failure
2 start of ramp up
3 end of ramp up

reliability owing to the increased component count, and the difficulty in monitoring and controlling the parallel components.

Thus, we finish up with a compromise: transistor assemblies are available which will switch, at 2000–5000 pulses/s, currents of up to 300 A (Fig. 6.13). With such devices we can obtain noise levels over the power ratings envisaged of not more than 60 dBA, which is lower than the noise level encountered in most computer rooms.

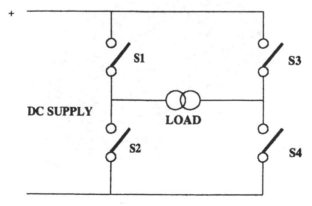

Figure 6.11 *Principle of inverter bridge*

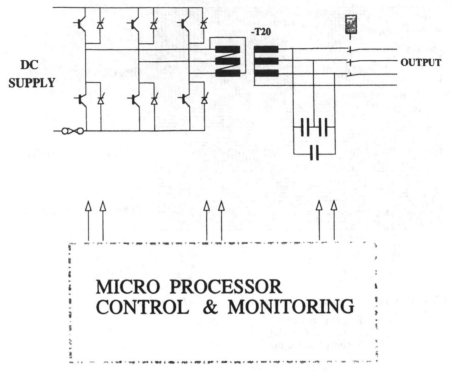

Figure 6.12   *Three-phase transistor inverter*

Figure 6.13   *Power transistor assembly — for switching 300 A at 2000–5000 pulses/s*

Each transistor assembly normally consists of two or three transistors connected in a Darlington configuration (Fig. 6.14), and there is often a diode within the circuit to assist in the rapid turn off of the device.

The basic inverter schematic is shown in Fig. 6.15; the three-phase output is achieved by displacing the inverter modules by the appropriate 120°. The square pulses generated are filtered, utilising the leakage inductance of the transformer and the capacitor filter to produce a sine wave. The high switching speed again results in an advantage in as much as the harmonics generated are also of a high order. This results in a small filter with low losses and low electrical inertia.

By producing a low-power very-accurate sine wave, we are able to compare the inverter output sine wave with this reference. The information regarding differences and deviations between the two waveforms is fed into a microprocessor, which in turn controls the main power transistors. The width of the square pulses is adjusted to compensate for any variations between the waveforms; the technique is known as pulse width modulation (PWM) (Fig. 6.15).

Fig. 6.16 indicates another advantage of the technique. For a 100% load application the voltage deviation is not more than 5%; it is, however, corrected within a few milliseconds. The response of the system is very fast, owing both to the switching speed and the low electrical inertia of the output filter.

We mentioned earlier the possibility of unbalanced loads together with some non-linear content. Three-phase monitoring and regulation is utilised within the inverter. In this way the voltage difference and angular deviation between phases is kept to the very minimum and is well within the limits set by computer manufacturers. With systems of this type, generally supplying a computer system comprising a number of devices, the non-linear content usually represents a small part of the load and is not a problem. It is however, always very difficult to guarantee there will be no increase in the distortion, unless we are able to measure the site and ascertain the harmonic content of the load.

Often, specifications quote crest factors. Whilst this is useful information in sizing the peak current requirements, if they do not tell us the duration of the peak, it is very difficult to ascertain whether the duration of the peak will significantly increase the RMS value.

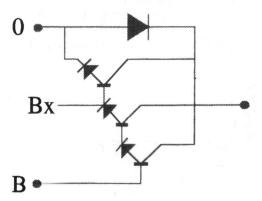

Figure 6.14   *Circuit diagram of power transistor*

Figure 6.15   *Principle of pulse width modulation*

UPS systems usually incorporate two forms of over-current protection: thermal detection for long slow overloads (Fig. 6.17), and electronic detection for fast high overloads. In either case the current is normally limited to 150% of nominal RMS, which corresponds to a peak of 150% × √2. The system would sustain this level for up to 60 s.

The output circuit is designed to have a low impedance at the fundamental frequency. At frequencies other than the fundamental the impedance will vary, as the nonlinear content of the load will draw current at the harmonic frequencies. This in turn will result in harmonic voltages being developed across the output impedance, and the value of the individual harmonic voltages will be:

Voltage harmonic = Impedance at H × Current H

where H = a particular harmonic frequency

The total voltage distortion rate can be calculated from the following:

$$\text{Voltage distortion rate} = \sqrt{\frac{V_{H2}^2 + V_{H3}^2 + V_{H4}^2}{H_1}} \text{ etc.}$$

Hence, it can be seen that it is a fallacy to assume, because it is transistorised inverter, that distortion will not occur, as is sometimes indicated. Regardless of the type of UPS, the harmonic voltages will appear across the output circuit. What we must do is design the output filter so that the effects of non-linear loads are minimised. What the transistor inverter does offer as an advantage is the ability to correct the voltage distortion much faster than either the thyristor or rotary UPS system. One of the most recent innovations in inverter technology is

100% load change

Figure 6.16   *Voltage deviation with 100% load change*

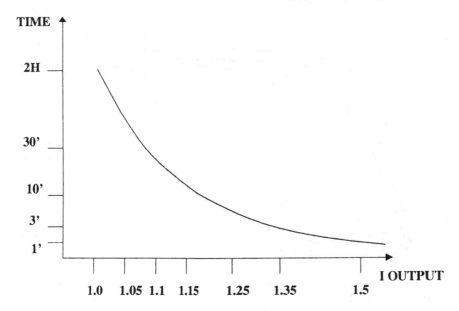

Figure 6.17 *Overload characteristics*

the use of Double PWM. A second set of control pulses are generated which are the inverse of the first set. The advantages are a much faster response, the output filter being even smaller which reduces the losses and, most importantly, the system's ability to cope with nonlinear loads. Typically systems incorporating Double PWM will have less than 8% distortion even with 100% nonlinear load application.

Another feature of transistor systems is their efficiency. As transistor devices approach saturation their efficiency tends to fall off very slightly, owing to hole storage effects. Unlike the thyristor, which operates most efficiently at 100% load, the transistor is at its best around 60–75% load. In practice this gives us a very flat efficiency curve for the transistor unit from 50 to 100% load (Fig. 6.18), with the efficiency at its maximum around 70%. As can be seen, whilst the full load efficiency is similar for all types of unit, it is only the transistor system which exhibits this flat characteristic. With all the other systems the efficiency decreases as the load decreases.

Most UPS systems are oversized for one reason or another. At the time of purchase it is envisaged there will be future expansion, or the rating is increased to cater for non-linear loads; it is very seldom that systems are fully loaded. Hence, when considering the efficiency of a system, it is well worth while looking at it when partially loaded. For example, at 5·5 pence per kWh, on an 80 kVA system loaded at 60 kVA, a reduction of 5% in the efficiency would increase the running costs by approximately £1700 per annum.

The advances in components and circuitry for the control of power devices have also helped to achieve these increases in efficiency, fortunately with no associated increase in the component count. The inclusion of microprocessors allows us to make software adjustments of the system parameters, often

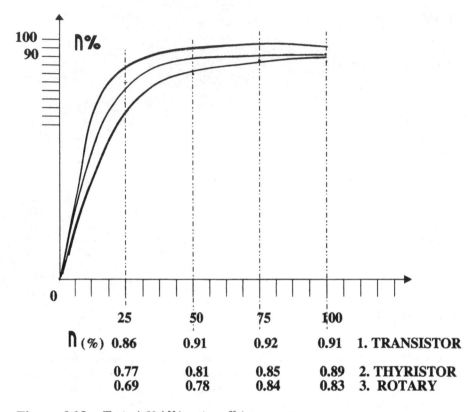

Figure 6.18   *Typical 60 kVA rating efficiency curves*

eliminating the necessity of on site potentiometer adjustments and the introduction of human error.

Many of the facilities which in the past have been considered as special or extra are now incorporated as standard; this again is due to the advances made in component technology. If one considers measurements, we now have a liquid-crystal display (Fig. 6.19) controlled by a touch sensitive selection pad, which can display input, output, DC voltages and currents at a reduced cost compared with that of the old analogue instruments and associated selector switches.

6.5 Static switch

The inclusion of a static switch is an absolute necessity for systems in this power range. It comprises a pair of 'back to back' thyristors in each phase of the bypass supply (Fig. 6.20) with a parallel contactor.

In the event of overload or failure of the inverter the load would be transferred from the UPS to the mains supply by means of the static switch. In order to achieve this, certain parameters must be met; obviously the two

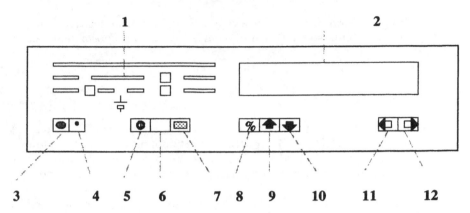

Figure 6.19   *Control unit*

1 alphanumerical display
2 buzzer stop
3 alarm light
4 scroll up
5 scroll down
6 numerical keypad
7 rectifier charger status

8 load on mains 2 indication
9 load on inverter indication
10 voltage measurement call
11 main menu call
12 current measurement call
13 keypad entry

sources, inverter and bypass, must be synchronised and the voltage must be within reasonable limit (±10%).

Under normal operating conditions the inverter will synchronise within the bypass supply, provided the bypass is within acceptable limits to the load; the tolerance is normally ±1% of nominal frequency (the tolerance range is, however, selectable over a range of ±0·5 to ±2%). To ensure no break transfer the phase must be within 3°.

If the transfer has been initiated by an overload condition, on the cessation of the overload the load would again be transferred to the UPS. In this manner the static switch would supply the inrush current normally associated with initial switch on, again avoiding the necessity to oversize the unit.

When selecting a system, check that the overload capability of the by-pass is greater than that of the inverter.

6.6 Diagnostics

A feature of these systems is the extensive diagnostics now incorporated. Again they offer several advantages at minimum cost, not least of which is the reduction in down time in the event of failure. The site operator can very easily and quickly fault-find down to subassembly level, using the display and control panel. The information passed to the service department enables them to arrive on site knowing which subassembly to replace and having it with them, thus avoiding the stocking or holding of spares on site.

Following the replacement of the faulty item, the system checks itself prior to allowing the switching-on procedure to continue.

An additional attraction of some systems is their continuous recording of parameters and events so that in the event of a fault the information can be accessed.

6.7 Communications

Various options are available for communicating with the UPS system. A monitor panel (Fig. 6.21) can be located remotely not only to give the status of the system but also to provide control. Software packages are available to display and control from a desk top microprocessor (Fig. 6.22), or to interface with the BMS system.

For communication over short distances (up to 100 m), the remote facility would be hard wired. Over greater distances the information would be relayed via modems and telephone lines.

Some manufacturers are now utilising the diagnostic and communication facilities to provide remote detailed diagnostics from their service headquarters. This will in the future probably result in a reduction in maintenance contract prices and possibly cessation of routine service calls altogether.

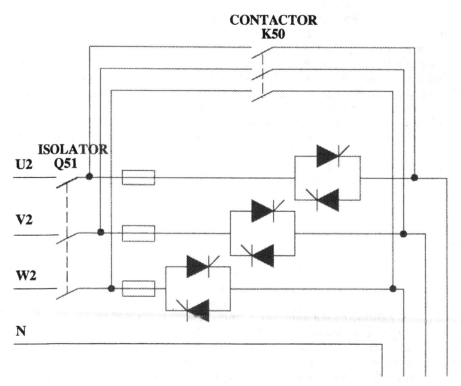

Figure 6.20   *Static switch: circuit diagram*

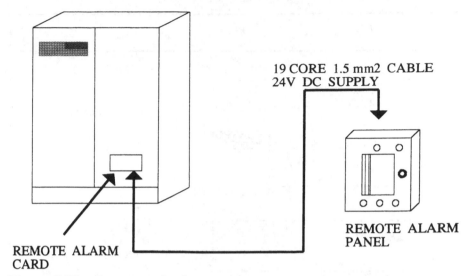

19 CORE 1.5 mm2 CABLE
24V DC SUPPLY

REMOTE ALARM
PANEL

REMOTE ALARM
CARD

Figure 6.21   *Remote control unit*

A number of the major computer manufacturers are now specifiying a requirement for communication between the UPS and the computer; again this is fitted as standard or at a small extra charge.

6.8 Construction

Ideally the subassemblies need to be plug-in or easily accessed to take full advantage of the diagnostic facilities, with the associated reduction in down time. With the exception of the very large power components (transformers,

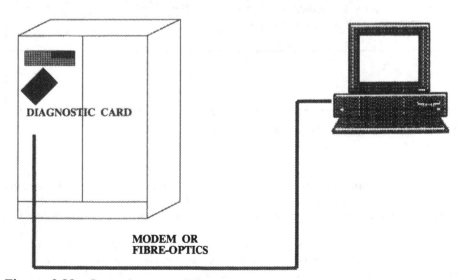

DIAGNOSTIC CARD

MODEM OR
FIBRE-OPTICS

Figure 6.22   *Personal computer link*

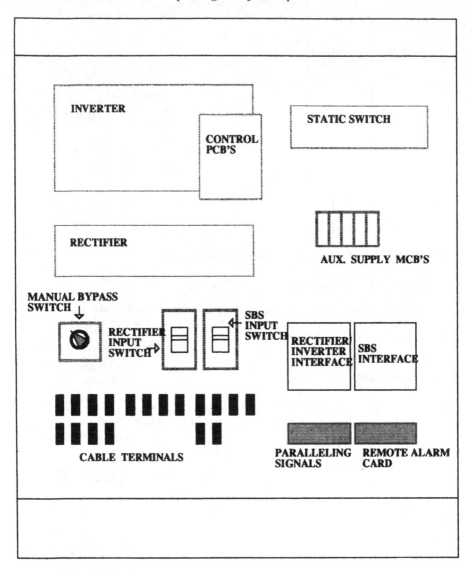

Figure 6.23   *Cubicle layout: rectifier-inverter cubicle 60 kVA*

isolators), the vast majority of the remaining components are located on either PC boards or the plug-in modules. The layout of the unit is such that they are accessible from the front as are in incoming and outgoing terminations (Fig. 6.23).

Fan assisted cooling is generally utilised: air is drawn into the bottom of the unit and exhausted from the top and back. Hence, clearance around the unit is required. Redundant fans are required and in the event of fan failure an alarm is indicated.

Figure 6.24a *Typical computer room UPS system of 120 kVA*

In some cases a special double skin construction is used, especially if the units are to be sited in a computer room, as it reduces considerably the possibility of airborne interference. Special filters are incorporated to eliminate conducted interference to the level stipulated by BS800.

Typical systems are shown in Fig. 6.24a and b.

6.9 Installation/environment

As has been shown, the systems incorporate control technology similar to that used in computer systems, and, in the ideal situation, they are installed in a similar environment. If this is the case the reliability is enhanced. As we are not temperature cycling the components, the battery is working within its ideal temperature range, again ensuring its life expectancy. The clean air-

Figure 6.24b   *The 120 kVA system with doors open, showing detail diagnostic information and safety panel*

conditioned environment reduces the possibility of dust build up within the system causing possible restriction of ventilation.

The units are however reasonably robust, and with adequate maintenance, the systems will work quite satisfactorily in other well ventilated reasonably clean environments.

From a practical point of view, attention must be paid to floor loading, input, output discrimination, isolation for maintenance and similar aspects.

The provision of separate supplies from the LV board, to both the charger and bypass input, helps to ensure input/output discrimination and usually provides a greater degree of isolation.

Chapter 7
Batteries

I. Harrison

7.1 Introduction

7.1.1 *Definition*

It is usually accepted that the definition of an uninterruptible power supply is one in which, in the event of the failure of the primary source of energy, a secondary source of power is available in such a manner that the load requiring the energy does not detect any break (measured in terms of microseconds) in the supply. Given this definition it will be found that nearly all the numerous standby power applications for batteries would fall into the category of an uninterrupted power supply system.

7.1.2 *Applications*

Examples of these would be: standby power for telecommunications network switching; power industry switch operation and switch tripping; and standby power for electronic equipment (usually computers) whether used for data-processing, information technology, industrial processing, or indeed hospital theatres.

Other applications, such as emergency lighting, alarms and sensors, are mostly used in a similar manner, though this type of application is not necessarily critically dependent on a 'no transient break' supply.

7.1.3 *Durations*

The standby power duration required by the applications mentioned can vary from consecutive bursts of high power of a few seconds each up to a more modest level of power over a period of time, say, between 3 and 10 hours. It so happens, however, that the standby power requirements for computers used in data-processing and information technology require power at rates of discharge between 5 and 15 minutes with occasional requirements going up to half an hour. Typically, emergency power is required in order that the computer system can be closed down safely, and without loss of database information or work in progress, at the time of the failure of the primary source of power.

7.1.4 *UPS*

Significantly, this last application has grown into a large industry and is still attracting a fairly high market growth. It has not only developed into a considerable industry but it has basically adopted the generic term (UPS) for its own use. It is this area of applications that this chapter on batteries will be addressing.

7.1.5 *UPS types*

Chapters 3–6 identify four different types of UPS system. Each system would conform to our general definition of UPS, but in fact is using the power available from a battery in three totally different ways.

The least significant from the battery point of view is the dynamic system with kinetic energy store described in Chapter 4. In this case the battery's only purpose is to start a diesel engine, and as such is not part of the UPS function and will not be considered further.

On the other hand, the dynamic system with battery energy store described in Chapter 3 is one in which the battery does provide the emergency power, but the dynamic part of the system does not require the battery to provide the 'transient no break' supply. In this particular case, when the primary source of energy fails the battery is actually switched into the system to provide the emergency back-up power.

In contrast the static inverter systems with battery energy store described in Chapters 5 and 6 are ones in which the battery does provide the 'transient no break' supply when the primary source of energy fails.

The dynamic and static types of system are completely different in the manner in which they use stored battery power. The implications of this difference in application will become manifest as this chapter on batteries develops.

7.2 Battery options

7.2.1 *Lead or alkaline*

The first choice within the battery options would be to choose between a lead–acid battery or an alkaline battery of the nickel–cadmium type. Whilst there is no doubt that the nickel–cadmium battery can give a good performance in the UPS system the cost of the cells and batteries (largely due to the costs of materials) make the nickel–cadmium cell an unattractive proposition. As a consequence of this situation, and because also the UPS business is highly competitive, very few nickel–cadmium batteries are used in UPS systems. For this reason, therefore, this chapter on batteries will concentrate on the lead–acid battery options that are available for UPS systems.

7.2.2 *Lead–acid batteries*

Lead–acid batteries appear in three generic types of construction. Each type reflects the type of positive plate used in the cell, and it is the type of plate which in turn determines the essential electrical characteristics of the product. The

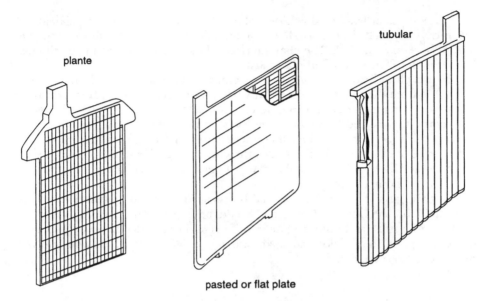

Figure 7.1 *Types of lead-acid battery plates*

three types of lead–acid cell available are (i) the Plante cell, (ii) the Pasted Plate cell and (iii) the Tubular cell. Illustrations of the three types of plate are shown in Fig. 7.1.

It is well known that all the types of lead–acid battery identified are not in practice used in UPS systems. The reason for this arises from the basic needs of the UPS application.

7.2.3 *Essential features*

The main criteria deciding the choice of a battery for a UPS system can be summarised in the following headings:

  (i) Discharge voltage profile
 (ii) Internal resistance
(iii) Energy density
(iv) Service life
 (v) Cost

Of the five features identified, all but (iv) service life, are related to the design and electrical discharge performance of the cell. The question of service life is more related to the electrochemistry of the cell and as such is dealt with separately at a later stage in the chapter.

Batteries for UPS applications are required to operate at the higher rates of discharge. Typical rates of discharge are from the 5 min rate up to the 30 min rate, with the most popular being a requirement for a standby period of 15 min. At these discharge rates there is a pronounced difference in the performance of the three types of lead–acid cell described. This variation, while dependent on the individual cell design, is shown schematically for the Plante cell and the Tubular cell in Fig. 7.2. The characteristic of the Pasted or Flat Plate cell falls between the Plante and Tubular cells.

The graph indicates the order of current as a function of the 10 h capacity available from the two types of cell at various rates of discharge. It is clear that at rates of discharge of a duration in excess of one hour there is little difference between either of the two types. However, at rates shorter than one hour the Plante cell's performance becomes increasingly more attractive until at the 5 min rate of discharge, it discharges the current equivalent to 2·3 times the 10 hour capacity, whereas a Tubular can only manage a current equivalent to 1·5 times its 10 hour capacity.

On this basis it could be reasonably assumed that the preferred choice of lead–acid battery for UPS systems would be firstly Plante, secondly Pasted Plate, and lastly the Tubular Plate cell. This selection order based upon purely electrical performance does not take into account the costs of the cells and the anticipated service life.

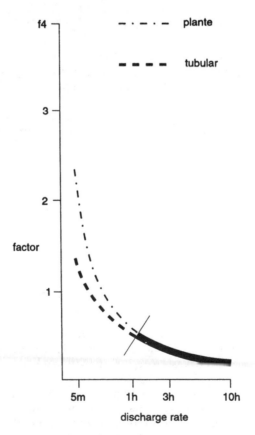

Figure 7.2   *High-rate performance comparison*

Table 7.1 *Cost comparison of lead–acid cells*

Type	Plante	Tubular	Pasted plate
Capacity (Ah)	400	400	400
Internal cost (1)	100	80	75
Life (years) (2)	20	14	12
Annual cost	5	5·7	6·3

(1) The cost figures are arbitary units with the Plante cell attributed with a value of 100.
(2) The life values assume operation in a true floating battery situation. The significance of this observation becomes clear when, later in the chapter, UPS systems are considered which do not operate the battery in a true float situation.

7.2.4 *Costs and service life*

In very general terms, comparison of costs between the three generic types of lead–acid cell is given in Table 7.1.

The principal point to observe here is that if costs are amortised over the expected life of the cell then the preferred order would be firstly the Plante cell, then the Tubular cell, and finally the Pasted Plate cell. In real life, however, it is the Pasted Plate type of cell which is the preferred product for most UPS battery systems. The reason for this is that service life requirements in UPS battery systems rarely exceed ten years, and a considerable number have a service life requirement of no more than five years. Under these circumstances amortising the cost of the cell over its expected service life is unimportant, and selection is determined by the product with the lowest cost.

7.2.5 *Valve regulated lead–acid batteries*

Having identified the Pasted Plate type of lead–acid cell as the optimum choice for most UPS systems, it is appropriate that some observations should be made on the most recent development of this generic type by considering the fairly recently introduced valve regulated type of lead–acid battery. Most of the valve regulated lead–acid batteries commercially available do use the Pasted Plate type of construction. That is not to say that the other types of plate could not be used in this kind of battery, but for reasons already identified the preferred choice of construction revolves around the Pasted Plate. The main feature that distinguishes the cell from a free venting conventional type of cell is the replacement of the vent plug on the conventional cell with a Bunsen type valve assembly on the valve regulated type of cell. A typical type of valve assembly is shown in Fig. 7.3. There are, of course, variations on the theme of this valve design but the one illustrated is typical of a very large number of products commercially available.

It should be noted that very often a valve regulated lead–acid cell is referred to as a sealed lead–acid cell. The use of the term valve regulated, however, is to attempt to draw attention to the fact that this type of product is not a hermetically sealed device, which is an implication that in some areas of operation the term could be taken as synonymous with the use of the word sealed.

The main feature of this type of battery is that it does not require additions of water to make good that lost as a result of electrolysis throughout its service life. The purpose of the valve is to encourage the recombination reaction within the cell, to reduce the losses of water as a result of electrolysis, and at the same time to release to the atmosphere small quantities of excess gas produced during normal operation, without at the same time allowing air from the ambient atmosphere to ingress back into the cell. This subject is dealt with in more detail later in the chapter when we consider the question of service life in terms of the internal electrochemical efficiencies of the cell.

It should be noted, however, that, in order to achieve the reduced loss of water by electrolysis, a recombination reaction is taking place and this in turn is brought about by cell design changes which in themselves have improved the discharge voltage profile of the cell, its internal resistance and energy density. Thus the product is not only desirable in UPS systems from the maintenance point of view but it also has an improved electrical performance in those areas critical to the UPS applications.

7.3 Battery characteristics

7.3.1 *The mystery box*

In many systems it appears that the battery is not only there to provide emergency power in the event of failure of the primary source of energy, but is also the panacea to all problems related to the regulation of the rectifier to the load. It is used as an infinitely variable source of electrical capacitance smoothing out irregularities on the DC power side of the system, and in so doing is generously given the title of 'soak away' or 'sump'. This feature is all taken for granted in the absence of any knowledge of what is actually happening inside the battery box and with little consideration as to the possible implications to

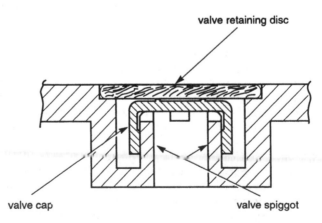

Figure 7.3 *Valve assembly*

service life. The purpose of this section is to give some indication in electrochemical terms as to what is actually happening inside the cell and from which explanations will be drawn when the operational systems are considered in battery terms in two types of UPS system considered in this book.

7.3.2 *Electrochemical efficiencies*

The discharge reaction in a lead–acid cell may be summarised as:

$$PbO_2 + Pb + 2H_2SO_4 \rightarrow 2PbSO_4 + 2H_2O$$

This means that both the lead dioxide of the positive plate and the sponge lead of the negative plate react with the acid in the electrolyte to form lead sulphate and water. The discharge reaction is electrochemically 100% efficient. This observation applies to both discharge through an external load or self discharge, (i.e. open circuit losses).

The recharge reaction in a lead–acid cell is the reverse of (i), and may be summarised as:

$$2PbSO_4 + 2H_2O \rightarrow PbO_2 + Pb + 2H_2SO_4$$

In this case lead sulphate on both the positive and negative plates reacts with water to form lead dioxide and sponge lead respectively on the positive plate and the negative plate with acid in the electrolyte.

The efficiency of the recharge action can be 100% but it is related to the state of charge of the battery or cell. When the recharge reaction becomes less than 100% efficient, the gassing reaction (electrolysis of water) takes place. This reaction may be summarised as:

$$2H_2O \rightarrow 2H_2 + O_2$$

In this case electrolysis of water takes place with an evolution of hydrogen from the negative plate and oxygen from the positive plate. The variation of recharge efficiency with the state of charge of a lead–acid battery is summarised in Table 7.2.

It should be noted that reference to the 80% state of charge is typical and not specific. The actual percentage state of charge at which the transition from 100% recharge efficiency takes place is determined by the design of the cell and the rate of discharge at which 100% state of charge (capacity) is defined. The important issue to note is that the transition does occur. From Table 7.2 it can be concluded that:

(*a*) Above 80% SOC, the required ratio of recharge in ampere-hours to discharge in ampere-hours is of the order 9:1.
(*b*) This ratio is capable of fully recharging a lead–acid battery provided the only discharge reaction (i) occurring between 80% SOC and the fully charged state is that arising from the self discharged process.
(*c*) If the reservation in (*b*) does not apply, then it is highly probable that the battery will not be able to recharge and never reach its fully charged state.

7.3.3 *Float voltage and open-circuit voltage*

The relationship between the float and open circuit voltage is important in so far as the float voltage is often fixed by the system, whereas the open circuit voltage is a function of the cell design. The difference between the two voltages

Table 7.2 *Electrochemical efficiency related to battery state of charge*

State of charge	Reactions	Electrochemical efficiency (% of current flow)
Maintenance in fully charged condition	(i) Discharge (self)	100% equivalent current
	(ii) Float recharge (compensation)	10%
	(iii) Electrolysis	90%
Approximately 80% state of charge	(i) Discharge	100%
	(ii) Recharge	100% at 80% SOC* – 10% at 100% SOC*
	(iii) Electrolysis	NIL at 80% SOC* – 90% at 100% SOC*
Fully discharged to approximately 80% state of charge	(i) Discharge	100%
	(ii) Recharge	100%

*SOC = 'state of charge'

is known as the overvoltage. This determines the value of the float current, which in turn affects the ability of the float current to maintain the cell in the fully charged condition.

If the cell is not maintained in the fully charged condition then compensation for the self discharge reaction (i) is not adequate, with the consequence that the cell will gradually 'walk down' in capacity until the cell reaches 80% SOC. This condition can be brought about by:

(*a*) The overvoltage being too small to sustain the correct float current. (It should be noted that this situation has in some situations been done deliberately to reduce loss of water, but to compensate for the resulting loss of capacity, regular 'freshening' charges are provided.)

(*b*) Systems regulation apparently meeting the overvoltage requirement, but depreciating the value of the float current (see Section 7.4 for examples).

7.3.4 *Resistive impedance*

Many 'users' of batteries regard them as the panacea for problems of DC regulation regardless of how the battery performs these functions and, in consequence, regardless of the possible adverse effect upon battery life and performance. Most of these effects arise from the factors already mentioned coupled with the variation of resistive impedance with state of charge.

Relative impedance levels at various states of charge are shown for a lead–acid battery in Table 7.3. Because actual impedance levels vary with the type and size of battery, the impedance values in Table 7.3 are based on a unitary value of 1 being taken for the lowest impedance reaction (i) at the 100% SOC.

The important feature to note is the considerable difference in impedance that exists between the recharge reaction (ii) and the discharge reaction (i). Fig. 7.4 shows this difference graphically.

7.3.5 *Typical floating battery operation*

In the simplest case of standby power operation the normal order of events would be:

(*a*) The battery would discharge and its impedance would increase according to reaction (i) in Fig. 7.4.
(*b*) Following the discharge period, the prime power source would be restored, and the battery would be recharged to the fully charged condition along the plots for reactions (ii) and (iii) in Fig. 7.4.

This situation is perfectly acceptable to the floating battery, but what would happen if the system in which the battery is floating is oscillating the battery between discharge and the recharge/float charge condition.

7.3.6 *Oscillating conditions on the DC line*

In the region 0–80% state of charge, both reactions (i) and (ii) of Fig. 7.4 are 100% efficient and therefore the state of charge would move in the path of the reaction carrying the larger current. However, in the region above 80% state of charge, a situation exists where recharge currents are required in the ratio 9:1 to the discharge current, in order that the capacity taken out of the cell can be restored.

Table 7.3 *Relative resistance impedance related to battery state of charge*

State of charge	Relative resistance Reactions	Impedance
Maintenance in fully charged condition	(i) Discharge (self) (ii) Float recharge (compensation) (iii) Electrolysis	>1<113))>113)
Approximately 80% state of charge to 100% state of charge	(i) Discharge (ii) Recharge (iii) Electrolysis	100% SOC* = 1 80% SOC* = 1·4)100% SOC* = 113)80% SOC* = 75
Fully discharged to approximately 80% state of charge	(i) Discharge (ii) Recharge	80% SOC* = 1·4 0% SOC* = 4 0% SOC* = 38 80% SOC* = 75

*SOC = 'state of charge'

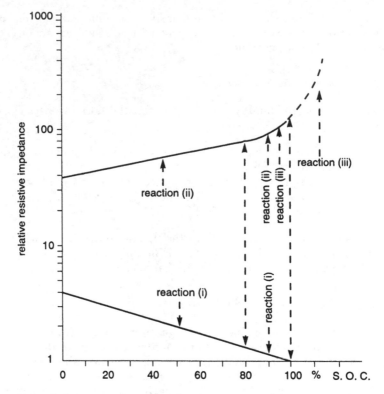

Figure 7.4   *Relative impedance related to battery state of charge*

7.3.7 *Detection*

It should also be noted that transient excursions into the discharge condition could be made frequently, but because of the low impedance of the discharge reaction (i), the excursion need not be obvious from voltage observations. Indeed probably the only way to reliably detect transient oscillations between the recharge and discharge reactions is to detect the direction of the current through the battery.

7.4 Battery operation in various systems

7.4.1 *Battery–rectifier systems*

This is the most simple battery system, and given a rectifier with adequate smoothing across its output, the battery would behave as a typical floating battery. However, if we were to consider the battery floating across a full wave rectifier without any 'smoothing' then the theoretical voltage profile would be that shown in Fig. 7.5.

This, of course, disregards the effect of the open circuit voltage of the battery, where, in the case of that part of the half wave voltage falling between $V_{o/c}$ and

V_{max}, the battery would receive a corresponding pulse of charge current. Where the half wave voltage falls below $V_{o/c}$ it could be argued that the battery may discharge back through the rectifier in an attempt to maintain the system voltage at as high a level as possible. This would not normally happen because the rectifier would present too high an impedance to accept a current out of the battery.

Whether a discharge pulse is taking place or not, it cannot be detected by observations of voltage across the battery. The natural voltage/time inertia of the battery reaction to change completely disguises the effect. The only means of detecting a discharge pulse through the battery is to actually record the direction and amplitude of the current.

The important issue is that the difference between $V_{o/c}$ and V_{max} must be large enough to make the amplitude and the width of the charge pulse of sufficient magnitude to ensure a net charge current capable of recharging the battery and maintaining it in the fully charged condition.

7.4.2 *DC ripple*

The subject discussed above has been known for many years under the description 'AC ripple', and the use of this term owes more to the fact that it arises from fluctuations in the unsmoothed rectified supply than it does to alternating current. As in fact there is no reversal in the direction of the current output of the rectifier, the condition is better described as DC ripple.

7.4.3 *Battery–rectifier resistive load system*

In this case we are considering a constant current load which in normal circumstances is powered by the rectifier in parallel with the battery. The difference between this situation and the simple system considered earlier is that the rectifier is now being asked to provide a much higher current into a relatively low impedance load, at a voltage at least between $V_{o/c}$ and V_{max} to protect the battery.

Again, given a rectifier with adequate smoothing across its output, the battery would behave as a typical floating battery. If, however, the rectifier is

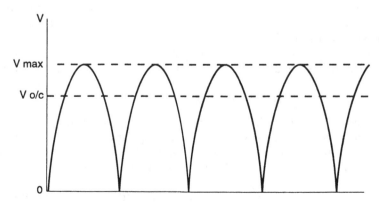

Figure 7.5 *Theoretical voltage/time profile*

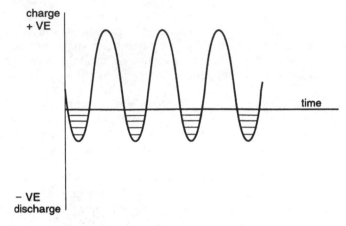

Figure 7.6   *High frequency shallow cycling (HFSC)*

inadequately regulated the opportunity is now far greater for the battery in its low impedance discharge mode to 'assist' the rectifier in smoothing the demand by the load. This condition is shown graphically in Fig. 7.6.

This situation can only be detected by recording the direction and amplitude of the current flowing through the battery.

7.4.4 *High frequency shallow cycling*

The subject described earlier is really high frequency shallow cycling, which is a condition where the battery is constantly subjected to an oscillatory charge/discharge mode by the system. It has been referred to as 'AC ripple' in the past, but in battery terms it is high frequency shallow cycling (HFSC).

7.4.5 *Consequences*

The consequences of this oscillatory condition will probably mean that the battery may never reach the top of charge condition, which in turn can be detected qualitatively by the observation of persistent float currents higher than the values quoted by the manufacturer. To avoid these consequences rectifier outputs should be adequately smoothed, and the float overcharge voltage should be sufficiently high to ensure the correct level of float current.

It should be noted that the output of a 3-phase rectifier is by nature smoother in output, and offers advantages in respect of the issues discussed earlier.

7.4.6 *Battery–rectifier pulsed load system*

In this system we are specifically addressing the type of load typical of some inverters that takes its DC power in the form of pulses. With this situation the ability of the rectifier to meet the pulse demand of the inverter is critical to the good operation of the battery. Failure to meet the pulsed demand will result in an acute case of HFSC.

The battery that would be most resistant to the effect of HFSC would be one with a relatively high discharge impedance, and operating with an extended overcharge voltage which in turn would attempt to keep the battery in the high-impedance recharge/gassing mode. Regretfully, the effects of the higher over-voltage would have almost the same effect upon battery service life as the HFSC that we are trying to avoid.

The solution to the problem is to have a design of rectifier capable of accommodating the variable pulse input of the load. This, however, adds expense to the system, and frequently the battery is used as a 'soak away' power regulatory device.

7.4.7 *Detection*

Again the condition can be detected and quantified by recording the duration and amplitude of the current passing through the battery. A more qualitive indication of the phenomenon can be obtained from the general level of float current. In some cases permanent float currents have been high enough to indicate that batteries very rarely exceed the 80% state of charge, and never receive a complete commissioning charge to bring them to the fully charged condition.

The implications to service life in this non-floating application are adverse and because the nature of the HFSC cannot be predicted it has so far proved impossible to make sensible life predictions for the batteries.

7.5 Battery sizing

All battery sizing techniques assume implicitly that, following a discharge, the battery system is capable of, and is given sufficient time, to fully recharge the battery according to the manufacturer's recommendations. For this reason we distinguish between the 'floating battery in standby applications', and the 'floating battery in a working application'. In the former application, and assuming an adequately designed float charge system, there is certainly sufficient time to recharge the battery following a discharge to the manufac-turer's recommendations. In the case of the working battery application, however, there is usually insufficient time to fully recharge the battery in a floating battery recharge situation. For this reason, the following subclauses deal with battery sizing separately to suit the needs of the two applications.

7.5.1 *Floating batteries in standby applications*

Typical standby applications would be telephone network switching batteries, most emergency power and alarm systems, and UPS systems which avoid high frequency shallow cycling. In all of these applications, and assuming an adequately designed rectifier, there is adequate time to fully recharge the battery, prior to its next discharge.

7.5.2 *Principal requirements*

The factors that affect the ultimate size of a battery can be summarised as follows:

(i) Power required

(ii) Voltage window (difference between the maximum charge voltage and the minimum discharge voltage)

(iii) Rates of recharge

(iv) Temperature

(v) Life

When a battery sizing calculation is complete the result is a statement of the number of cells required to do the duty, and the ampere-hour capacity required of each cell; the ampere-hour capacity, of course, always being referred to a specified rate of discharge defined by time and end voltage.

The number of the cells in the battery is normally determined by the maximum rate of recharge required in conjunction with the upper limit of the voltage window. The ampere-hour capacity of the cell in the battery on the other hand is determined by the required power, the lower limit of the voltage window, ambient temperatures, and a factor to allow for the depreciation of actual capacity at the end of life.

7.5.3 *Constant current battery sizing*

Traditionally, battery companies have supplied data using a form of engineering note by which the size of a cell in the battery can be calculated from the current required, the duration of that current and the end of discharge voltage. There are many techniques for calculating the size of a battery: all tend to work on the same principles, the only differences being matters of detail in the technique. One technique is to use battery calculation curves normally generated by the battery manufacturing company. A typical set of curves is shown in Fig. 7.7.

Very simply, the capacity of the cell required can be calculated by taking the specified load current and multiplying it by the factor obtained from Fig. 7.7. For example, consider a requirement for 160 A for 5 min to an end voltage of 1·90 V per cell. Refer to Fig. 7.7, and drop a vertical line from the 5 min curve

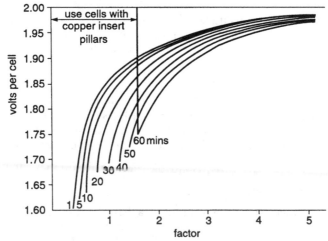

Figure 7.7 *Battery calculation curves*

Constant current discharge performance data											
Discharge currents (amperes per cell) at 20°C to 1.75 V per cell											
Cell type	Minutes							Hours			
	5	10	15					1	1.5		

Figure 7.8 *Battery calculation tables (format)*

at 1·90 V per cell, and this will give a factor of 1·85. A product of the load times the factor gives a required capacity of 296 Ah.

This calculated cell capacity would then be amended to suit the required temperature of operation. Most battery sizes are rated to an ampere-hour capacity at 20°C, and if the operating temperature is above this value then there could be a case to reduce the required capacity; on the other hand if the operating temperature is below 20°C, then it is highly likely that the cell capacity will have to be increased in value. The data necessary to perform this function are usually obtained from manufacturer's documentation. To accommodate depreciation of capacity during service life, it is necessary again to refer to manufacturer's data. Typically, with Plante cells, it is not necessary, for the type of construction of the cell allows it to give 100% of claimed capacity throughout its service life. All other types of lead–acid cell have service lives quoted with a depreciation in claimed capacity. Typically, the end of service life capacity would be of the order of 80% of that claimed.

Whilst the battery curve method of calculating the size of a cell in the battery is still used by battery engineers, it is gradually falling from use with system engineers, and tables of data containing standby time to various end voltages are preferred. An example of this is shown in Fig. 7.8.

The systems engineer will have at his disposal a family of tabulated data giving cell performance at incremental voltages between 1·9 V per cell and 1·6 V per cell. It should be noted that similar tables for sizing a battery on a constant power basis are also available and, of course, are used in an identical manner to that used for looking up constant current performance values.

7.5.4 *Floating batteries in working applications*

Typical working battery applications are: switch operation in the power industry, solar power, and load stabilisation in UPS system. In all these applications sizing of the battery to suit the duty follows the same route as for calculating the battery size in a floating battery in a standby application. In the working battery application, however, the main criterion to establish is whether the battery can be brought to the fully charged state, and for how long.

In the case of the switch operation application, batteries are given regular freshening charges to keep them in the peak condition. In the case of solar power, however, the recharging cycle is fairly predictable but infinitely variable in output. The net consequence of this situation is that systems engineers appear to size a battery for the worst condition in anticipation of loss of capacity during service, and base service life on their own system predictions. Similarly, static UPS systems, which display high frequency shallow cycling, appear to be managed in a similar manner. The degree of battery oversizing is probably governed more by the competitive situation than it is from a systems operation situation.

7.6 Battery standards and safety codes of practice

In closing this chapter on batteries, reference must be made to national product standards and safety codes of practice. There are a series of national British standards which cover the various types of battery that can be used in various stationary battery applications. These standards contain information which differentiates the characteristics of each different type of cell and provides test methods by which comparisons can be made. Other EEC countries, notably France and Germany, have similar suites of standards, and harmonisation of all these standards into European norms will be an equally useful source of information. The recommendation is: use the standards.

There are many authorities dealing with the subject of safety of electrical installation. Batteries however, being a unique component in DC systems, have attracted little interest from regulatory organisations. However, there is a code of practice available in the form of a British national standard, which is a document that is continually being updated to reflect the current situation, and is a very useful consultative document. Again, codes of practice of this type will be harmonised with other EEC standards. Its objectives are very similar to the European counterparts, the differences being only the rate at which individual countries can publish and adopt new measures in anticipation of the European harmonisation of all standards.

Chapter 8
Applications to air transportation

R. CATO

8.1 Introduction

This chapter will review the application of uninterruptible power supply systems (UPS) used at airports, by airlines and by air traffic control. In the UK, Heathrow, Gatwick, Stansted, Prestwick, Glasgow, Edinburgh and Aberdeen are operated by BAA. Manchester, Birmingham and Newcastle Airports are operated by PLC companies, a number of other airports are operated by the local authority and the small airports in Scotland are owned and operated by Highlands and Islands Ltd. The on route Air Traffic Control facilities are operated by National Air Traffic Control Services, which is a joint service provided by the Civil Aviation Authority and the Ministry of Defence.

There were two significant changes recently in the Industry during 1986, namely the privatisation of British Airways and BAA. Both of these organisations employ professional engineers to design and maintain their electrical distribution systems and safety is, of course, of paramount importance. Safety remains as a top priority and it is significant that, despite the increased commercial pressures resulting from privatisation, there has been little cultural change in the electrical distribution philosophy used in these organisations. If anything, the increased commercial pressure has resulted in the need to improve the integrity of the power supplies to protect commercially sensitive data.

It will become apparent that within the UK there is only a limited requirement for UPSs for most airport functions. However, there is a growing demand for UPSs for airline computer reservation systems and there is a continuing demand for substantial UPSs for air traffic control functions.

8.2 Regulatory framework

The Air Navigation Order requires most flights in the United Kingdom for the public transport of passengers, and all flights for the purpose of flying instruction, to take place at a licensed aerodrome or at a Government aerodrome or at an aerodrome managed by the Civil Aviation Authority. The Order also makes provision for an applicant to be granted an aerodrome licence subject to such conditions as the Authority thinks fit.

The conditions concerning the licensing of aerodromes are published in a document known as CAP 168 and it contains a reference to the requirements for power supplies to aerodrome lighting systems. CAP 168 contains the criteria which are given in the Standards and Recommended Practices, Annexe 14, Convention of International Civil Aviation. The power supply requirements for the navigational aids, which are mainly the responsibility of the CAA, are contained in Annexe 10 of the Convention.

At all international airports much depends, of course, upon the prevailing weather conditions and the runway visual range (RVR). Each runway at an airport is categorised with the following definitions:

Category I: Intended for operations down to 60 m decision height and down an RVR of the order of 800 m
Category II: Intended for operations down to 30 m decision height and down to an RVR of the order of 400 m
Category III
A: Intended for operations down to an RVR of the order of 200 m
B: Intended for operations down to an RVR of the order of 50 m
C: Intended for operations without reliance on external visual reference

CAP 168 defines the lighting services to be provided for each runway, the brilliancy settings and the minimum switchover time to the secondary power supply.

In general terms a 15 s break is acceptable for Category I operations, but most services require no more than a 1 s break in supply for Category II and Category III operations.

8.3 Airport power requirements

At most major airports there is a correlation between the annual number of passengers handled and the power supply requirements. The correlation changes slightly where substantial aircraft maintenance is undertaken. In the UK the correlation is:

$$1 \text{ MVA} = 1 \text{ MPPA (million passengers per annum)}$$

At Heathrow the load varies between 40 and 50 MW depending upon the time of day and season, and the annual passengers throughput for the year 1988–89 was approximately 38 million. It will be seen that the total UPS requirement for airports like Heathrow is a fraction of the total load.

The HV distribution diagram is shown in Fig. 0.1. Power is supplied to the airport via four main intake substations (three at 22 kV and one at 33 kV) and is distributed around the airport at 11 kV. Virtually all of the distribution relies on the closed ring main principle and the substations on the airfield lighting ring can be switched to the West Intake automatically should the North Intake completely fail.

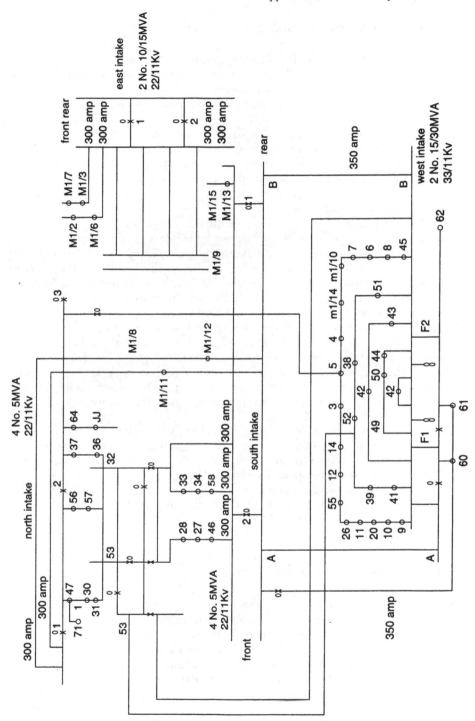

Figure 8.1 *HV distribution diagram*

8.4 Airfield lighting and control systems

The requirements of CAP 168 dictate that under Category I operations the airfield lighting services may be interrupted for a period of not more than 15 s. This can easily be accomplished by the provision of local diesel generating sets. Fig. 8.2 shows a typical lighting pattern under low visibility conditions.

Typically there are 800 lighting fittings on a Category III licensed runway supplied on 30 separate series circuits. Each light is fed by its own isolating transformer and the arrangement therefore overcomes a voltage drop and lamp failure. Each series circuit is fed by a constant current regulator and the series circuits emanate from four different substations, two at each end of the runway. The series circuits are 'interleaved' so that a failure of one circuit would still result in a recognisable lighting pattern.

Under Category II operations conditions the switchover time for the secondary power supply must not exceed 1 s. Unless special dispensation has been received from the Licensing Authority this means that in practice the diesel generators become the primary source of the supply and the mains becomes the secondary source of supply. At Heathrow, Gatwick and Stansted the supply is deemed to be so secure to the airfield lighting substations that the diesels are not normally started in the revertive running mode.

An uninterruptible service of the airfield lighting systems is therefore achieved by the application of simple distribution techniques rather than by a discrete UPS. In order to control the airfield lighting systems at busy international airports, however, UPSs are required to back-up the computer systems. At Heathrow the total number of airfield lighting fittings is approximately 10 000 and in order to maintain control (but not illumination) of these fittings UPSs are required.

The taxiway system at Heathrow is extremely complex, and with over 70 take-offs or landings an hour it is necessary to provide a control system which will guide aircraft around the airport and help them to taxi between the runway and parking stands. The ground movement control system (GMC) is a hot-standby dual computer system which permits the Air Traffic Controllers in the Control Tower to set up a taxiing route for an individual aircraft. A simplified block diagram is shown in Fig. 8.3.

8.5 Telecommunications systems

At BAA Airports a central telephone exchange provides an essential information service for the general public. In addition, for the airlines and concessionaires, the facility provides an on-site network which avoids BT network call charges. Typically at Heathrow there are over 3000 extensions connecting together 150 companies. At the smaller airports the equipment is centralised, but at the larger airports a distributed architecture has been adopted to overcome transmission losses and access line tariffs. To comply with the requirements for direct dialling facilities, at each of the locations where the equipment is distributed, a standby battery facility is provided. In some locations diesel generators are also installed.

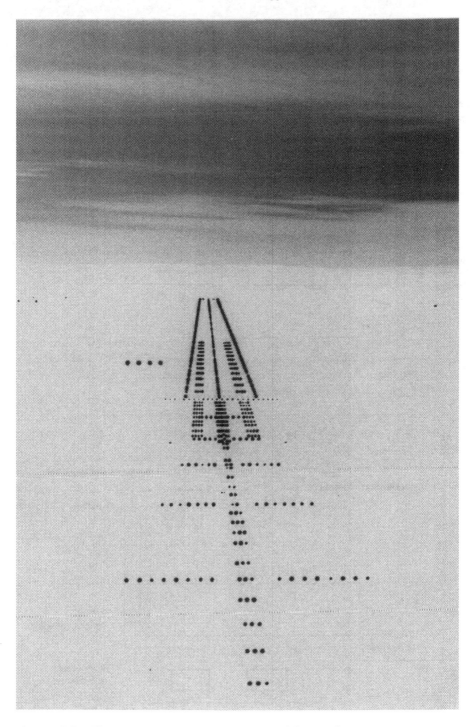

Figure 8.2 *Typical lighting pattern under low visibility conditions*

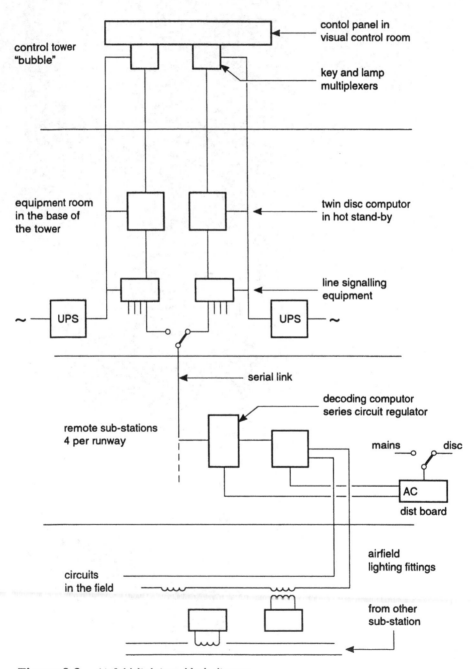

Figure 8.3 *Airfield lighting: block diagram*

Virtually all of the system exists therefore without the need for UPSs. There is, however, one aspect of the system which is more important than the transmission of voice traffic and that is the storage of the call charges incurred by the subscribers. At Heathrow the total traffic is in excess of 70 000 calls per day and it is therefore vital to capture the information and ensure that it is safe.

It has already been explained that the design of the external electrical distribution system mitigates the need to provide UPSs to the airfield lighting services and at some airports the internal distribution philosophy is such that there is no need to provide UPSs to the call loggers. At other airports less attention has been paid to the internal electrical distribution philosophy and uninterruptible power supplies of sufficient size are therefore required merely to support the call loggers. The power supply requirement for these systems is very low, typically 1 kVA, with the most onerous part of the duty associated with starting the Winchester disc.

8.6 Computer information systems

With over 60 million passengers passing through the South-East airports in 1988 and with over 540 000 air transport movements in the same period there is obviously a considerable amount of data held in airport flight information systems and airline reservation systems.

At Gatwick there is slightly less flexibility with the HV network compared with Heathrow and not quite so many options exist for feeding the terminal buildings. The view is therefore taken that the flight information system, which is a distributed computer system driving flap boards and television displays, should be backed up by UPS. At Heathrow, where the HV network is more flexible and where there are of the order of 800 television monitors used, no UPSs are provided.

The data held in the reservations systems, however, is vital to the continuing operation of the Airport, and British Airways, for example, make considerable use of UPSs. In 1984 Holec Ltd. installed a 5-set 450 kVA system in BA's Comet House Data Centre and in 1985 the company installed a 5-set system in BA's Boadicea House. At Comet House Data Centre the computer systems are responsible for reservations, departure control, fares and ticketing, hotels, message switching and flight planning. At Boadicea House Computer Centre the systems support the airline's accounting, engineering, crew rostering and personnel functions. In 1988 a further five 500 kVA diesel type rotary no-break UPSs were commissioned to complement the installation at Comet House. The equipment was commissioned to coincide with a major upgrade to BA's computer systems and to meet the data storage demand which was growing at a rate in excess of 30% per annum.

8.7 Airport navigational aids

As mentioned earlier the on-site navigational aids, at BAA Airports at least, are maintained by the CAA and comply with the requirements of Annexe 10 to the Convention of International Civil Aviation. In addition to the VHF radio systems the navigational aids include Distance Measuring Equipment (DME) and Instrument Landing Systems (ILS).

The ILS generates two radio beams which are used by the aircraft when they are on their final approach. The instruments in the cockpit enable the pilot to lock the aircraft on to an approach path which is not only on the extended centre line of the runway but which is at the exact descent angle to ensure the aircraft lands in the touchdown zone (TDZ) area of the runway in all weather conditions. The ILS equipment battery supply must, in order to meet the international regulations, be capable of running ILS for at least two hours.

The CAA's basic philosophy for all these systems is simplicity. The planned maintenance programme for the ILS, and in fact all equipment using these systems, requires periodic checking on load to ensure that the system does work in the event of outside electricity supply failure. The batteries used on the ILS are the modern seal type cell and they require no maintenance.

8.8 National air traffic control

It will be seen so far that at most airports in the UK limited use is made of UPSs for other than the protection of data and that the electrical distribution is so designed to reduce the requirement for UPSs. However, safety is not compromised and in many cases the installations far exceed the requirements of the international recommendations. The philosophy at the London Air Traffic Control Centre (LATCC) at West Drayton, however, is to make extensive use of UPSs and standby generators. The UK is divided into two Flight Information Regions, one of which is LATCC the other being Scottish Air Traffic Control Centre (SATCC) at Prestwick. Fig. 8.4 is a block diagram of the UK air space and shows the division of responsibility between LATCC and SATCC.

LATCC is responsible for providing air traffic services to all aircraft flying in or over the London Flight Information Region — England up to 55N, Wales, the Isle of Man, Northern Ireland and the surrounding seas up to the air space of adjacent countries. It is supported by a sub-centre at Manchester Airport which handles aircraft flying below 15 500 ft at Manchester and Liverpool Airports and in certain parts of the airways around Manchester and the Irish Sea.

SATCC is responsible for an area from 55N to 61N near the Faroe Islands. Eastwards, it reached 250 miles to the Norwegian and Danish Flight Information Regions. To the west, it extends about 150 miles to the Oceanic Control Area, taking in air space over Northern Ireland.

The site at LATCC is owned by the Property Services Agency (PSA) and virtually no reliance is placed upon the integrity of the HV supplies from the regional electricity supply company.

There is, of course, an 11 kV distribution system fed by two incoming feeders at 22 kV. There are a number of substations which are fed directly from the distribution system which are for PSA related activities, but the substations which are directly associated with LATCC functions are fed from that part of the distribution network which is powered by diesel generators and which uses the PSA distribution network as back up. Located within the generation house are six units rated at 1·5 MVA and two units rated at 0·5 MVA. The normal practice is to run three sets on load, which leaves adequate spare capacity for expansion and breakdown. Waste-heat recovery principles are used to increase efficiency of the generation system.

At the heart of the distribution system are three UPSs. The first system is a four-module Emerson UPS, each module rated at 250 kVA, 50 Hz. The second system is a four-module Siemens, each module rated at 330 kVA, 50 Hz. The third system is a two-module Siemens system, each module rated at 330 kVA, 60 Hz, and is used to drive the most important system at LATCC, namely the IBM 9020 computer.

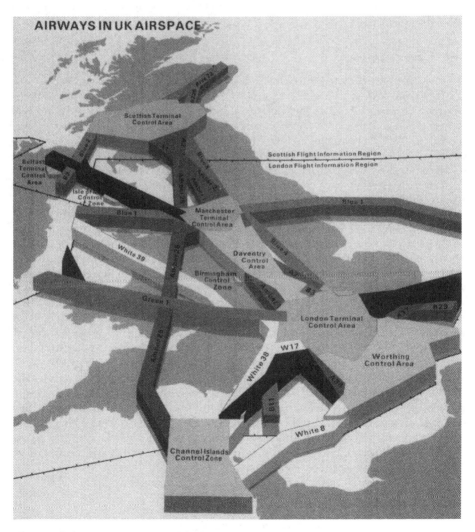

Figure 8.4 *Block diagram of UK air space*

Three HV feeds are taken from the HV network supplied by the generator house. After step-down these feeds are supplied to the main distribution board which is broken into three sections. Each of the 50 Hz UPS is connected to each of the three parts of the main distribution board. Each part of the main distribution board has a sub main distribution board for non-essential services.

The final main distribution board (FNDB) is split into two halves. Each half has two feeds from one UPS and one feed from the other UPS. There are additional connections from the non-essential sub main distribution boards. The final switching arrangement for the essential services permits a feed from either a UPS or a non-essential distribution board.

It is possible therefore to feed an essential service from one of two UPSs or directly from one of three HV/LV transformers supplied by the generator house, which in the event of total failure can be supplied by the local supply company.

8.9 Conclusions

When this chapter was being drafted, two interesting events occurred.

Firstly on 24th May during a heavy thunderstorm, a lightning strike hit the Central Terminal Area at Heathrow causing various computers to trip out and damage to a number of external closed-circuit TV cameras. At 1329 h a further strike hit an SEB transformer on the south side of the Airport. A dip in the power resulted which caused disruption to Terminals 3 and 4. A number of motor alternator sets supplying the 400 Hz for the fixed electrical ground power system also tripped out. By 1600 h all of the computers had been reset, and although there were some flight delays, there were no cancellations.

The other interesting event happened at Gatwick on Friday 16th June at approximately 1542 h. Although Gatwick has two intake substations only one of these is normally connected to the HV ring. That afternoon one of the two South Eastern Electricity Board's transformers at Smallfield Heath had been isolated for maintenance when the other developed a fault. When the supplies to the other Intake were energised further problems were encountered and it was not until 1715 h when a restricted load of 8 MVA could be re-established. During the period when there was no power available from SEEB the diesel generators were available and the UPSs protected the vital areas. Although there were one or two flight diversions the airport handled 72 flights in the period between 1542 h and 1752 h when all the supplies were re-established.

Incidents like the two at Heathrow and Gatwick happen very rarely and there is no pattern. Records are, of course, kept but the control engineers at Heathrow could not recall a similar incident on the distribution system within the last five years. Although the affects of the incidents were alarming the disruption was short lived and safety was not compromised. The conclusion therefore must be that the distribution philosophy is correct and that sensible use is being made of UPSs.

8.10 Acknowledgments

Grateful thanks are owed to the following: R. Plant, Civil Aviation Authority; C. Raper, National Air Traffic Control Services; S. J. Guttridge, Holec Ltd.

Chapter 9

Applications to telecommunications

N. G. M. NEWTON

9.1 Introduction

In the not too distant past the telephone only appeared on isolated desks, was used infrequently, prone to wrong numbers, plagued by noisy lines, subject to operators listening into conversations and thought of as an expensive message medium. If the telephone was out of order it was considered inconvenient. Today the telephone is provided on a one-to-one basis, it is in frequent use and, with the exception of mobile telephones, more accurate (most dialling errors are human) and clearer in audio reproduction, with no unwelcome listening in; also the costs for making calls, by comparison, are more economic. Now if the telephone is out of order it can be a major handicap.

In a relative short period of time the telephone has become an essential business tool expected to function uninterruptedly any hour of the day or night. There has therefore been a related growth in, and user demands for, emergency power support to telecoms. It is estimated that the European UK market for UPSs in 1991 was of the order £80 million. The major users are banking and manufacturing with telecoms coming a close third. In the United States telecoms accounted for 66% of the total market in 1988.

9.2 Uninterruptible power supply

For telephone systems to be operational at all times their power services must cater for any loss in mains supply. Additionally most modern telephone equipment incorporates computers and microprocessors which are susceptible to power interruptions of only a few milliseconds, and to relatively minor corruptions of the supply voltage, frequency and quality. Where the system is not already served from a battery a UPS will overcome these problems to provide a clean, constant source of AC power.

Solid-state inverter based static systems account for the majority of UPS units installed to protect communications equipment and networks. The continuing development in electronic technology is permitting manufacturers to make parallel improvements to UPS plant. Reduced size, improved efficiency, cool running, low noise output, stability and reliability are paramount.

A static on line UPS consists of four basic modules:

 (i) Phase controlled rectifier
(ii) Batteries

(iii) DC to AC inverter

(iv) Bypass/transfer switch

The phase controlled rectifier draws power from the AC mains, keeping the battery bank charged and holding the batteries at float voltage level. A filter between the rectifier and the batteries reduces ripple to maximise cell life. The battery will invariably be of the lead–acid, sealed, low-maintenance type (which actually still requires maintenance) with generally a 10 year operational life.

The inverter converts the DC power into AC, controlling voltage, frequency, harmonic distortion and power rating. This is normally achieved by pulse width modulation (PWM). The regulated AC output is fed directly to the power input terminals of the protected plant.

Should the mains power fail the inverter draws DC power from the battery and the load will operate without interruption until the battery discharges to a level defined by the battery characteristics and efficient operating parameters of the protected load. This time period must be carefully specified, together with the output load. A battery capacity to serve the exchange for 4–7 hours should meet the requirements of most emergency situations. When the battery reaches the critical discharge level the control will automatically switch the UPS into bypass even though the supply has not been restored.

When considering the mains power requirements to the UPS due allowance must be made for the losses in the rectifier and inverter, the applied load and the additional demand during the recharge cycle. During recharge the UPS support time will be less than specified and after a prolonged emergency can take 48–60 hours for the battery to reach peak condition. In the event that the UPS itself fails, the static bypass switch will connect the load without break to an alternative bypass power source, usually the mains input to the UPS. The ultimate choice of which type of UPS to use can vary with circumstances according to the equipment being supported and end user preference.

9.3 National networks

When considering the support of any equipment or circuits by UPS it is necessary to establish the risk — the risk of the AC mains failing or standards falling below acceptable tolerances and the knock on consequences of such a risk. Such ramifications may incorporate inconvenience, delay, cost; client or in-house dependence. A chain is as strong as the weakest link. In any network it is therefore prudent to aim for equal level of risk to all components in that chain.

The telecoms manager has control of the system destiny when planning and implementing internal networks. The situation dramatically changes when information is to be passed to a distant location. In such circumstances the user generally becomes at the mercy of the public telephone switched network (PTSN) operator. It is therefore necessary to assess the risks of failures on the PTSN to assure all links in the telecoms network are of similar dependability and, where not, to incorporate emergency measures to provide equality.

The major network providers take integrity of supply as a corner stone of their service. British Telecom from its Post Office Telephone origins has always provided at all main exchanges both battery and standby generator support. Most main exchanges operate on a 48 V DC supply off batteries, the battery being charged by local company mains in healthy state and standby generator in mains failure. The standby generator and UPS are carefully co-ordinated to ensure operational integrity.

As the networks have grown and the drive for cost effectiveness has become a commercial necessity, so the operation of outgoing remote nodes without backup has been introduced. A main host exchange interfaces between local exchanges and the main network; if the host fails then only local calls can be connected on the local exchange. In such circumstances duplication of essential circuits should connect to different host exchanges. The network providers, for obvious reasons, are most reluctant to identify or quantify the extent of such exposed operation; however, in all city and major conurbation areas it can be taken for granted that full standby and alternative automatic routing options are provided as standard. Total integrity cannot be guaranteed but a very impressive mean time between failures (MTBF) ratio can be assured.

9.4 Subscribers' equipment

With assurances on the availability of incoming lines, what of the private exchange equipment in the users premises? The regulations for maintaining services during power failure situations are not over generous. The general recommendation is for 20% of exchange lines to be operational in 'fall back' situations. Note this refers to exchange lines and not handsets on the system. In most cases it is necessary to change the MF handset in these defined locations to loop disconnect phones, before the lines may be utilised in a power failure mode. The operator's console is bypassed and non-functional during a power failure. Where direct dialling in (DDI) facilities are installed a 7 hour standby provision for all lines must be incorporated, where connecting to the BT network.

With the reliability of local electricity supply company services so high, the average subscriber gives little thought to mains failures and relatively few expend monies on full standby support of telephone systems. In too many situations the UPS is a grudge purchase added as an after thought, usually following bitter experience of an actual mains failure incident.

In public services, hospitals, security establishments, military, government etc., full standby backup is a specification requirement. Careful thought is given to ensuring total dependability of the electrical services to the telephone exchange with no-break operation. In critical areas the UPS switch-mode rectifiers are operated in parallel to provide the total power with an inherent measure of redundancy and an extremely high system MTBF.

In banking dealing rooms where voice, video and data services to the desk are paramount to successful business trading, the utmost reliability of electrical services is critical. These systems are usually linked to a main UPS serving all essential elements.

Large telephone exchanges invariably incorporate duplication of components with 'hot standby' of central processor and configuration to localise any minor failures. The equipment manufacturers have taken great care to ensure services are maintained wherever reasonably possible on plant or component failure. On small private automatic branch exchanges (PABXs) or key systems where economy is the major factor such duplication is rarely provided unless specifically called for by the purchaser. It should be noted that approximately 80% of all exchanges are under 100 extensions.

9.5 Effect of mains failure

It is worth considering what actually occurs when the electricity mains service to a telephone exchange fails.

The latest generation of telephone exchanges are microprocessor controlled, driven by a software package. The software on large telephone exchanges is recorded on disc which holds the information to 'boot up' when full power is restored. Systems now offer a multitude of features which the extension user can individually program. When the power supply fails these extension programmed facilities can be lost as systems do not generally enter this information on the master disc. In addition, recent extension changes entered via the telephonist's console can be lost if the master disc has not been backed up. It can take hours of reprogramming following a mains failure to reintroduce the *status quo* on large systems which have not followed a strict back-up routine.

Certain feature-rich handsets incorporate individual dry cells to maintain local information. The regular replacement of cells should be verified with the system maintainer.

On small key systems the memory is usually held on read only memories (ROMs) with in-built power units to hold integrity but not dive the system. The key system telephone units are invariably unable to operate on fall-back mode. To utilise exchange lines it is necessary to change the key system telephone for a traditional loop disconnect handset. In mains failure mode calls cannot be transferred as the mains system is inoperative. Only a few defined extension locations are usable; all others are de-energised.

9.6 Emergency power services

For the telephone user who is unable to link their exchange to a main UPS or where no UPS exists in the building, it is necessary to provide a dedicated unit.

Large PABXs and the public telephone network actually operate on DC. Generally each cabinet forming the telephone exchange contains a main service module which then converts the electric input to the voltage required by the individual shelf units with final DC/DC convertors at board level which produce the required voltage from 5 V to 24 V. These include units with 5 V and 12 V conversions to interface between analogue or external signal levels and logic circuits common in telecom equipment.

The manufacturer will configure the distribution within the exchange to meet the electricity service arrangement. If AC, the cabinet will contain a rectifier with various DC voltage outputs to meet the shelf unit requirements. Savings can be made serving an exchange direct with DC. Conventional UPS systems incorporate rectifiers to convert the battery power to AC. Standby power systems for large PABXs are manufactured specifically without the output rectifier to serve the equipment at DC. It is essential to consult the telephone exchange manufacturer/supplier when considering emergency support power services.

Telephone exchanges are not hungry for power; even for the larger units approaching 1000 extensions, electrical services of 32 A 415 V AC, are rarely exceeded. For the smaller end of the market, under 100 extensions, a 13 A single phase fused connection unit will suffice. For small telephone exchanges it is economic to use one of the many available micro UPS units primarily manufactured for stand-alone computer plant. Such UPSs are small in size and can be located alongside the telephone exchange in the office area. The latest generation of micro UPSs (under 500 VA) are of pleasing appearance and do not generate excessive heat.

It is recommended that automatic bypass be specified together with alarms and status indication. The small PABX/key system manufacturer should be consulted to ensure the output power and waveform are compatible with the system requirements.

Other factors to consider are safety and noise. IP31 is sufficient to protect the unit from hazards such as spilt coffee and paper clips. Sealed 'maintenance free' batteries by definition emit no obnoxious vapours. An acceptable noise level is 45 dBA at 1 m; if in excess of 60 dBA it would be subject to complaints from those working in the vicinity. The smaller UPS units are generally offered in 10–15 min versions and it is necessary to specify extension modules to increase the support time.

9.7 British Standards: Telecommunication

In general, the British Standards applying to the power services associated with telecommunications are concerned with the safety requirements. Their intention is to reduce adequately the risk of hazardous electrical conditions arising on the telecommunication networks and of electrical hazards for persons using the power supply unit. In a secondary capacity the standards cover the prevention of damage to property and other elements.

BS 6506, under clause 5.3 on power supplies or batteries, indicates it is general practice for standby power services to be rechargeable battery. It recommends their accommodation should comply with BS 6133 which relates to the safe operation of lead–acid stationary cells and batteries. BS 6484 says little other than to refer to BS 6301; and BS 6301 says a little more but is mainly concerned with safety.

British Telecom network exchanges require convertors to operate from generators with BS 649 class A2 governing, nominal output 48 to 58 V, design and construction to BS 5406 and Electricity Council Engineering recommendation G5/3.

Where telecom power supplies operate on the national network, they are required to conform with British Telecommunications Technical Guide no. 26. Smoothing of the output has to meet CCITT requirements, and ripple voltage must not exceed:

2 mV RMS psophometrically weighted to 800 Hz.
50 mV RMS between 45 and 125 Hz.
5 mV RMS above 3 kHz.

All when floating across a battery of ampere-hour capacity four times the current rating of the charger.

The current British Standards with direct application to telecoms are as follows:

BS 6301: Specification for safety requirements for apparatus for connection to BT networks

BS 6305: Specification for general requirements for apparatus for connection to the BT public switched telephone network

BS 6312: Specification for plugs for use with BT line jack units

BS 6317: Specification for simple telephones for connection to the public switched telephone network run by certain telecommunication operators

BS 6320: Specification for modems for connection to the BT public switched telephone network

BS 6328: Apparatus for private circuits run by certain public telecommunication operators

BS 6450: Private branch exchanges for connection to the public switched telephone network

BS 6484: Specification for safety requirements for independent power supply units for indirect connection to BT networks

BS 6506: Code of practice for installation of private branch exchanges for connection to the BT public switched telephone network

BS 6701: Code of practice for installation of apparatus intended for connection to certain telecommunication systems

BS 6789: Apparatus with one or more functions for connection to certain public switched telephone networks

BS 6833: Apparatus using cordless attachments (excluding cellular radio apparatus) for connection to analogue interfaces of public switched telephone networks

BS CP 1022: Code of practice for the selection and accommodation of telephone, telegraph and data communication installations

9.8 Associated telecom services

In the early 1980s the only item of consequence connected to the telecom network other than the telephone was telex machines. In the late 1980s the telex has been joined, and substantially overtaken in numbers, by the fax machine. The fax has become an everyday item appearing on office desks: it may not have legal recognition but this has not detracted from the speed, convenience and economic advantages. Where one fax per office existed in 1986, three now operate full stretch, side by side, with split for outgoing only and receipt.

The only cloud on the horizon is the disturbing growth in unsolicited sales literature which 'clogs up' the machine where more urgent necessary faxes await dial tone. Fax directories may be a thing of the past with users going ex-directory and only issuing numbers to those closely involved in their daily business. The government has promised legislation to empower Oftel to serve preventive orders on junk fax senders. A condition of the Branch System General Licence introduced by DTI, following advice by Oftel, allows recipients to take out injunctions or private legal actions.

Fax and telex machines derive power for the printing drive from the building electrical service. It is therefore necessary to make emergency power available if the continuing use of these facilities is required in electrical mains failure incident. The more sophisticated the services offered by the machines the more facilities will be programmable by the user; if one loses the power then the abbreviated dialling unique to the user, and other similar features, may be lost and require reprogramming. Top of the range units may, however, incorporate dry cells to protect integrity but not drive the machine. Fax machines rarely take in excess of 150 VA and telex units 200 VA.

Where equipment cannot be interfaced with main-building UPS and standby generator protected circuits, local micro UPS units can offer an economic and simple solution requiring no major cabling works. Some units incorporate standard 13 A, 240 V, 50 Hz socket outlets to plug in the apparatus to be protected, and flex fitted with 13 A plugs to serve the UPS in healthy electrical mains status.

In dealing rooms and financial institutions there has been a mushroom growth in information services with video and, recently, digital feeds to the desk positions. Large equipment rooms now house the service cabinets, each unit generally requiring a 13 A 240 V, 50 Hz feed. Protection of the electricity supply by UPS and standby generator is the norm throughout the major London dealing room operators. Smaller organisations are now becoming ever more aware of the information services on offer and the benefits to be on stream. The vendors have therefore developed equipment to serve a single desk or a small group which can be located in the telecom intake, comms. room or beside the desks served. These isolated items of kit can go unnoticed when compiling the schedules of plant requiring emergency support. Where installed in buildings with no main generator or UPS a small stand-alone UPS can be considered.

Information services are not confined to financial institutions; large successful networks have been established nationwide to all types of business from travel agents for booking holidays, flights etc. to legal professions with all case law available on screen with frequent update on latest judgments. Each circumstance must be assimilated: to the travel agent it is critical to business and must be operational at all times, so UPS support becomes a specification requirement. It is generally the user's responsibility to provide the UPS.

9.9 Data/telecom interface

The telephone network was initially established worldwide for the purpose of voice communication; if any other medium wished to utilise these service links, the operating parameters had to abide by the rules and requirements dictated by the voice transmission. When computers first required to speak to distant cousins they were forced to use voice circuits and accept their severe limitations.

How times change — the merger of voice and data on a digital fibre optic nationwide telephone network is now taken for granted by the new generation of communications engineer. We are but one small step away with fibre distributed data interface (FDDI) from linking the national network through an in-house combined network to the desk position. But that is the future: what of today?

Today to make more efficient use of the nationwide links multiplexers and statistical multiplexers interface between the in-house computer operations and the external public telephone network. These become ever more clever with automatic alternative routing should there be line failure. The multiplexers themselves require a local 13 A, 240 V, 50 Hz, power service. If communications are to continue during in-house mains failure incidents then local UPS support must be provided.

For main computer operations the provision of a UPS with standby generator will invariably be the norm. The risk is at the distant location — a small branch, shop or supplier where only the most basic of computer services exist and little if any emergency power backup. The loss of a day's trading information can be obviated by interconnection at all distant telecom interface points of individual UPS support units. Where modems are used infrequently they can be configured through the telephone exchange to use the general outgoing exchange lines, which offers assurance of high dependence. Single dedicated lines can always fail, but the incidence of all switchboard lines failing is more remote.

Electronic mail has not been a success; it has only low penetration and the client base is generally in small specialised areas. The interface between the national telephone network and the in-house equipment should be supported by a UPS. Each of the network operators' systems are not identical, some can store and forward messages when the mail box becomes free, others cannot; in such circumstances a power failure can be an imposition on both sender and receiver.

There are approximately 5000 British Telecom Kilostream users, and feedback from telecom managers indicates that the initial teething problems have been overcome and this 64 k bit/s service is now offering a reliable network connection. The connection at each end of the national link within the users premises is by a network termination unit which requires a low-power local service. This local power service will require UPS support if the continued operation of the link is necessary during mains failure incidents.

The foregoing generally applies to Megastream links and the comparative services as offered by Mercury. Where using British Telecom Packet Switch Stream, care must be taken to ensure that adequate emergency backup is incorporated as the network operator only provides battery backup in key areas.

The support of interface equipment applies at all levels from the basic modem to the ultra-modern electro-optic drivers and receivers.

9.10 Installation and environmental factors

UPS units give off heat owing to inefficiencies of the rectifier and inverter. The telecoms equipment itself will also raise room temperatures. Attention must be given to adequate ventilation and the introduction of cooling plant in rooms containing large units.

It is wise to restrict water services entering or passing through the equipment room. Where unavoidable, the provision of water detection units under the raised floor with remote alarm at a manned station will quickly instigate evasive action. Avoid basement rooms with drain gullies or manhole covers because, in storm conditions or blockage, water can back up and flood the UPS equipment.

UPS units are particularly heavy owing to the lead–acid batteries. The structural support capability of the raised floor must be checked. It may be necessary to mount the UPS on an angle frame over a spreader plate direct onto the structural slab, and butt the raised floor up to the support frame.

Verify the practicalities of the proposed delivery route through the building. Will the UPS fit within the goods hoist? Can the platform take the full weight or must the UPS be delivered in component form for fabrication in the final position? Are access doors wide enough? Is a crane required for off loading and lifting to roof? Is it necessary to remove a window frame or section of external cladding to obtain direct entry to the proposed room?

The UPS does not strictly have to be located in the communications room: can it be accommodated in a car park space with partition enclosure? Any remote site must offer an interconnection route for service control and alarm cabling.

As the UPS is a critical item, is a Halon fire discharge protection required? If so, the room must be sealed with automatic control interconnections to vent the plant and Halon-extract system. Interlinks with the main fire alarm system must activate the evacuation/alert procedures.

Check that the UPS quotations include all necessary associated electrical interconnections and switchgear to make the system operable.

Remote alarms of mains failure must give warning to users that they are on emergency services with only limited resources. Certain UPS manufacturers incorporate output facilities for interconnections to a building management system. The status and alarm signals can then be presented at the main building control centre with the assurance of structured procedures being actioned.

It may be that after a set period, if the mains supply re-connection looks to be some time too distant, an instruction is issued to restrict telecom traffic to essential use only.

The power taken by exchanges is in direct relationship to the traffic passing through — reduce the number of calls and the support duration is extended. A 4 hour battery backup will extend to 24 or even 48 hours with light usage; e.g. at weekends. Alarms can be incorporated to signal set reductions in the output voltage from the cells to indicate remaining capacity.

Ensure there is emergency lighting to the room containing the UPS so that, in mains failure situations, staff can check on the unit's condition.

The equipment should be designed to operate within the electrical noise limits, both psophometric and EMI, required for communication environments. Radio frequency interference should be in accordance with the requirements of VDE 8071.

For a large UPS consideration should be given to the installation of an independent earth connection back to the main electrical intake or sub-station. A UPS cleans the power downstream but pollutes the service up stream owing to the thyristor control sequencing. Frequency distortions back fed are usually of the third and fifth harmonics.

When connecting a UPS unit to a standby generator the plant must be compatible, with the generator oversized to accommodate the idiosyncrasies of the UPS thyristor control. It is essential both manufacturers are brought together prior to placing of orders to verify the inclusion of all necessary special provisions. Particular attention must be given to the engine governor and the over/undervoltage and frequency detector systems.

9.11 Maintenance

A full maintenance agreement is recommended to ensure any UPS unit installed is regularly monitored to confirm integrity of operation. A further factor is the inclusion of the emergency power module within the switchboard maintenance contract under a single responsibility.

A repair by replacement maintenance philosophy should be considered which permits the removal of faulty sections and direct replacement with tested and operational units. The equipment can be immediately brought back on line with the faulty units repaired off site.

9.12 The future

With the large scale integration of voice and data distribution, sharing of cable and management medium, data nationwide networking via the telephone public service, the time may come when one wonders why they were ever considered as separate.

Developments are proceeding where the emergency power unit is fitted on the individual cards to be supported. Micro power sources integrated on the chip are a reality and used by AT & T in very specialist applications. As the technology becomes more efficient and cost effective, self support without the fitment of mains UPS may perhaps become the norm for telecoms.

The introduction of fibre distributed data interface (FDDI) and the inevitable acceptance by the telephone network providers of utilisation of a dedicated channel within the 100 Mbit/s bandwidth will make the installation of a structured fibre optic cabling distribution system within high-technology buildings the norm. The telephone system would be connected to the spine cable with nodes per floor or department for star burst service to the individual telephones.

The recent developments in compression of voice, to pack more channels into less bandwidth, will continue to reduce the numbers of cable interconnections and equipment, thus reducing the overall requirements for standby power.

The termination of a fibre optic cabling medium direct to the handset and the true digital–digital interlink are just over the horizon but may take some time yet to become the economic commercial reality.

Currently telecoms require emergency services for two elements — the exchange and the feature handset. The future will add a further element, the support equipment for the in-house cabling medium.

Chapter 10

Harmonic distortion of UPS input and output voltages

A. C. KING

10.1 General note on non-linear loads and distorted currents

A non-linear load may be defined as a load which, having a sinusoidal voltage applied to it, passes a non-sinusoidal current. Many everyday loads are non-linear but the non-linearity is often unimportant: an unloaded transformer is non-linear owing to magnetic saturation. Significant non-linear loads associated with UPS equipments include rectifiers and switched mode power supplies.

Any regular non-sinusoidal waveform may be regarded as a compound wave made up from a fundamental component and harmonic components: the composition of such waveforms may be established mathematically by applying Fourier's analysis. As triplen harmonic currents have zero phase sequence they cannot exist in a three-phase balanced form unless there is a neutral return path. Even harmonics indicate a lack of symmetry between positive and negative half cycles.

The phase sequence of three phase harmonics is in accordance with the following pattern:

Harmonic number	1	2	3	4	5	6	7	8	9	10	11	12
Phase sequence	+	−	0	+	−	0	+	−	0	+	−	0

The rectifiers normally included in UPS equipments are of the three-phase bridge type; as there is no neutral connection there will be no significant triplen harmonic currents. As positive and negative half cycles should be balanced no significant even harmonic currents are expected. Any that are present will be due to a lack of symmetry between the firing of pairs of thyristors. The half-controlled bridge generates even harmonics but it is not usually found in UPS equipments and will not be further discussed.

The 'power factor' of a non-linear load can have several meanings; it follows that whenever the term is used it should be defined. Assuming a pure sine wave of voltage, only the fundamental current component can supply power, any harmonic components being parasitic. However, industrial ammeters normally indicate either the RMS or the rectified average value, both of which include the harmonic components. To calculate power from current and voltage readings it is necessary to include a distortion factor μ in addition to the conventional power factor, $\cos \phi$, of the fundamental current and voltage components.

Conventionally, non-linear loads are regarded as loads which take current at the fundamental frequency and which include, in series, harmonic voltage generators having zero impedance.

There are a number of formulae likely to be encountered in connection with non-linear loads; they appear below and use the following convention:

a = RMS value of the total current

a_1 = RMS value of the fundamental current component

a_n = RMS value of the nth harmonic current component

RMS value of total current = a

$$= \sqrt{a_1^2 + a_2^2 + a_3^2 + a_n^2} \text{ etc.} \tag{10.1}$$

Distortion factor μ = $\dfrac{\text{RMS value of the fundamental current component}}{\text{RMS value of the total current}}$

$$= \frac{a_1}{\sqrt{a_1^2 + a_2^2 + a_3^2} + \text{etc.}} \tag{10.2}$$

Total harmonic distortion = $\dfrac{\text{RMS value of harmonic components}}{\text{RMS value of the fundamental component}}$

$$= \frac{\sqrt{a_2^2 + a_3^2 + a_4^2} + \text{etc.}}{a_1} \tag{10.3}$$

Various formulae are used for this quantity; that given above is used in document G5/3 described in Section 10.3

Peak or crest factor = $\dfrac{\text{Instantaneous peak value of current}}{\text{RMS value of the total current}}$

$$= \frac{\text{Instantaneous peak value of current}}{\sqrt{a_1^2 + a_2^2 + a_3^2} + \text{etc.}} \tag{10.4}$$

For a sinusoidal supply voltage, and where $\cos \phi$ is the power factor of the fundamental current component:

$$\text{Power} = V a_1 \cos \phi$$

$$= V a \mu \cos \phi \tag{10.5}$$

The power factor, $\cos \phi$, of a diode rectifier is close to unity whereas that of a thyristor rectifier is determined by the commutation delay angle such that

$$\cos \phi = \frac{\text{Output DC voltage with phase control}}{\text{Maximum DC voltage with no delay}} \tag{10.6}$$

which can result in unexpectedly low power factors for lightly loaded rectifier equipments.

Typical values of μ and $\cos \phi$ are:

	μ	$\cos \phi$
Diode rectifier	0·96	0·98
Thyristor rectifier	0·94	0·5–0·9
Switched mode power supply	0·66	0·94

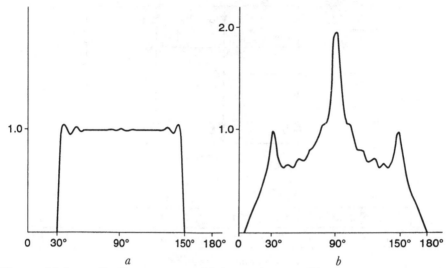

a   *b*

Figure 10.1   *a Quasi-square wave with harmonics up to 31st*
b The quasi-square wave of Fig. 10.1a with reversed 5, 11, 17, 23 and 29th harmonics

The profile of a non-sinusoidal quantity is determined by the magnitudes and phase relationships of its harmonic components. It follows that the profile of a distorted current cannot be determined from the distortion factor of the total harmonic distortion alone; hence the use of the peak or crest factor.

The currents indicated by Figs 10.1*a* and *b* both have the harmonic components (from 5th to 31st) characteristic of a six-pulse rectifier. For Fig. 10.1*b* the signs or 'polarities' of the 5th, 11th, 17th, 23rd and 29th harmonics have been reversed to achieve the high current peak. For Fig. 10.1*a* the addition of the infinite number of higher characteristic harmonics would result in a waveform having vertical edges and a flat top.

The approximate RMS magnitudes of the harmonics of both waveforms are:

n	1	5	7	11	13	17	19	23	25	29	31
RMS	0·78	0·156	0·11	0·07	0·06	0·046	0·04	0·034	0·031	0·027	0·025

Data relating to the two waveforms may be calculated:

	Quasi square wave	Peaky wave
Fundamental RMS a_1	0·780	0·780
Total RMS a	0·813	0·813
Distortion factor	0·959	0·959
Total harmonic distortion	0·294	0·294
Peak or crest factor	1·244	2·404

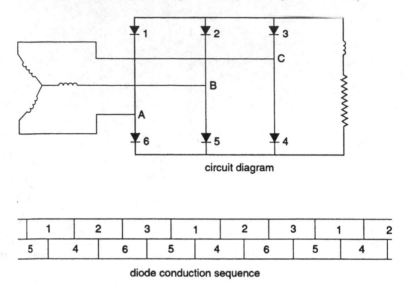

circuit diagram

diode conduction sequence

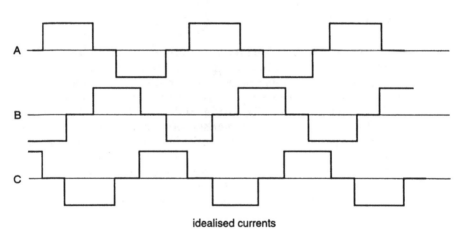

idealised currents

Figure 10.2   *Operating sequence and idealised currents of a three phase bridge rectifier*

10.2 Rectifier generated harmonics

10.2.1 *Rectifier operation*

The UPS equipments described in Chapters 3, 5 and 6 take power from the mains supply to feed a rectifier input module. The usual rectifier configuration is a three-phase bridge, which theoretically can demand a current with a harmonic content up to 30% of the fundamental. These harmonic currents, in flowing through the impedances of the supply system, cause distortion of the supply voltage for local consumers.

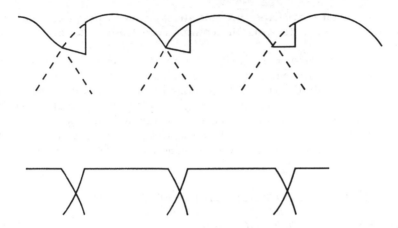

Figure 10.3   *Effect of diode commutation on voltage and current*

A three-phase bridge rectifier operates as two complementary star groups, each arm conducting for 120° in each cycle and the two groups being displaced by 180°. The operating sequence of the arms and the idealised current waveforms are indicated in Fig. 10.2.

For a three-phase bridge the harmonic content of such idealised square waveforms includes only 5th and 7th, 11th and 13th, 17th and 19th etc. The magnitude of these currents is always proportional to the reciprocal of the harmonic number, the 5th for example is 20% of the fundamental and the 13th is 7·7% of the fundamental. The total harmonic content (RMS value of all the harmonics) is 30% of the fundamental.

10.2.2 *Commutation*

The rectangular waveforms of Fig. 10.2 would require an instantaneous transfer of current from the conducting phase to the next in sequence. Such instantaneous transfer cannot occur owing to the presence of inductance in the supply; in practical circuits commutation between diodes requires a time of the order of 1 ms. Commutation has the effect of slightly reducing the higher harmonic content of the line current.

During commutation the direct current is supplied from two diodes connected to different transformer phases. As both diodes are conducting and connected to a common point the transformer phases may be considered as being connected together. The common point assumes a voltage that is the mean of the two phases until commutation is complete. Fig. 10.3 illustrates the effect of diode commutation on the transformer voltage and the DC current output.

With a diode rectifier commutation occurs at practically zero voltage difference, and rates of change of currents are not high. If the rectifier is of the controlled type it will incorporate thyristors instead of diodes, and reduction of the DC output voltage will be achieved by delaying the commutation. During delayed commutation the voltage difference between the phases can be high, which leads to a high rate of change of current and a short commutation period.

The idealised current shapes are not affected, but in practice, since the load inductance is finite, delaying the commutation leads to a change of current shape.

As in the case of diodes, during commutation the voltage of the common point assumes a voltage that is the mean of the two phases, leading to a voltage dip in the succeeding phase and a voltage rise in the preceding phase. At the end of the period the previously conducting thyristor abruptly assumes a reversed bias state. If the supply inductance is small it may be necessary to add series inductance to limit the commutating di/dt. Fig. 10.4 illustrates the effect of delayed commutation on the supply voltage and the DC current output.

10.3 Effect of a rectifier on a supply system

10.3.1 *Distortion of the supply voltage*

A rectifier may be regarded as a load which takes power from the supply system at fundamental frequency and generates harmonic currents which are fed back into the system. In flowing through the system the harmonic currents cause voltage drops along the various current paths, and these harmonic voltages distort the system voltage waveform. The distortion will be greatest at the rectifier terminals and the degree of penetration into the system will depend upon the system impedance. If a number of rectifiers feed harmonic currents into different points of a system the effect is cumulative, the penetration can be deep, and a large number of consumers may be affected.

In 1976, with this in mind, the Electricity Council issued Engineering Recommendation G5/3: 'Limits for harmonics in the United Kingdom Electricity Supply System'. Following the privatisation of the Area Electricity Boards in 1990, the Electricity Association is now responsible for Engineering Recommendation G5/3 which continues to be invoked by the distribution companies as the document relating to distortion. The maximum size of rectifier

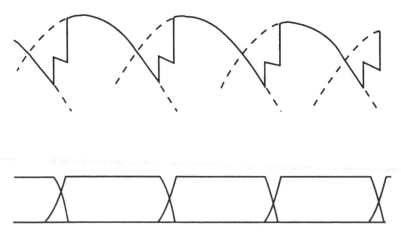

Figure 10.4   *Effect of delayed commutation on voltage and current*

allowed by this Recommendation depends upon the supply voltage and the pulse number, and is given in three stages:

Stage 1: Small equipments for which no detailed consideration of harmonic distortion is necessary before connection.
Stage 2: Larger equipments which may be allowed if certain criteria are met.
Stage 3: Large equipments requiring a detailed examination of existing harmonic voltages and current conditions, and of the new conditions that will result from the load.

A calculation of the effect of harmonic currents on a supply system is complex and requires a knowledge of the resistance and reactance of the various components such as overhead lines, cables and transformers. The voltage distortion produced at the point of common coupling between a rectifier and the supply may be estimated by assuming the supply impedance to be purely inductive:

$$\frac{\text{Harmonic voltage}}{\text{System voltage}} = \frac{n \times I_n}{I_{sc}} \tag{10.7}$$

where n = harmonic order

I_n = harmonic current

I_{sc} = symmetrical prospective fault level

This is a rearrangement of the expression given in Engineering Recommendation G5/3 Clause A 3.6.1. Any resistive loads connected to the system provide additional shunt paths for harmonic currents and reduce the harmonic voltage distortion. This simplified approach ignores the presence of capacitance which may exist as cable capacitance or as power factor correction capacitors.

Voltage distortion of the supply system is undesirable for a number of reasons, some of which are mentioned in the following paragraphs. If a rectifier is fed from a relatively small local generator such as may occur under standby power conditions, the voltage distortion will be larger than that experienced on a low impedance public supply system. The subject is discussed in Section 10.4.

10.3.2 *Power factor correction capacitors*

Any voltage distortion appearing across a power factor correction capacitor will cause harmonic currents to flow. The currents will be proportional to both the harmonic voltages and the harmonic numbers; they can be large and can damage the capacitors, particularly if the supply is being taken from a local generator.

Any lumped capacitance in the system may resonate at a harmonic frequency with transformer leakage reactance. Resonance causes large harmonic currents to flow in the components, and can lead to large harmonic voltages in parts of the system. Fig. 10.5 indicates the manner in which series and parallel resonances may arise.

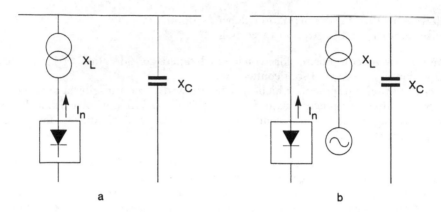

Figure 10.5   *Examples of series and parallel arrangements of transformer leakage reactance and a capacitive reactance*

a X_L and X_C in series
b X_L and X_C in parallel

The manner in which resonance is likely to arise is indicated by the following consideration of a 1 MVA 50 Hz transformer with leakage reactance of 0·05 p.u. and a 254 V (phase to neutral) secondary winding:

$$\text{Leakage reactance} = \frac{\text{volts per phase} \times \text{p.u. leakage reactance}}{\text{rated current}} \quad (10.8)$$

$$= \frac{254 \times 0 \cdot 05}{1312} = 9 \cdot 68 \text{ m}\Omega \text{ per phase}$$

Leakage inductance $= 30 \cdot 8 \, \mu\text{H}$ per phase

Resonance is unlikely to occur at the 5th harmonic and the 7th will therefore be considered:

For resonance at 7th harmonic,

$$\text{Capacitance} = \frac{1}{(2\pi f)^2 \times L} \quad (10.9)$$

$$= \frac{10^6}{(2 \times \pi \times 350)^2 \times 30 \cdot 8}$$

$$= 6700 \, \mu\text{F per phase}$$

At 254 V and 50 Hz the capacitor current would be 535 A, and when operating as a three-phase bank the kVAr rating would be 407 kVAr.

A 400 kVAr capacitor bank would be used, for instance, to correct the power factor of a 1 MVA load from 0·7 to 0·95 and is therefore likely to be encountered in practice.

Similar calculations undertaken for the 11th and 13th harmonics lead to resonant capacitor ratings of 165 and 118 kVAr, respectively. Such lower ratings may well be encountered if a capacitor bank is automatically switched in sections. A calculation for the 5th harmonic leads to a resonant capacitor rating of 798 kVAr, which would be unlikely to occur in practice.

Capacitors can be detuned by adding sufficient series inductance to ensure that the series circuit is inductive at the lowest significant harmonic (or the troublesome harmonic). The circuit will therefore be inductive at all higher harmonics.

10.3.3 *Other effects*

Data and control circuits running parallel to circuits carrying rectifier currents may be affected by inductive coupling, particularly if the rectifier uses thyristors. Telephone circuits may similarly be affected; the human ear is particularly sensitive to frequencies of about 800 Hz, the 16th harmonic of 50 Hz. A three-phase bridge rectifier produces the 17th harmonic.

Within asynchronous induction motors harmonic stator currents cause flux systems which rotate at harmonic speeds and cause additional iron losses, particularly in the rotor. Within synchronous machines the effects of harmonic currents and commutation are discussed in Section 10.4.

Some electronic devices use the supply voltage waveform for timing purposes. If the waveform is not sinusoidal the zero points or cross-over points will not occur when expected to do so and performance may be impaired.

Moving coil indicating instruments measure the rectified average of AC values but are usually calibrated in RMS values using a form factor of 1·11. If the waveform is distorted the form factor will differ from 1·11 and the indicated values will not be correct.

Moving-iron indicating meters will indicate true RMS values, including any harmonic components. If the waveform is distorted the true RMS values will differ from the RMS values of the fundamental. With both moving coil and moving iron instruments confusion can be caused if power is calculated without including the distortion factor μ.

Induction disc type integrating meters are susceptible to error when connected to a severely distorted supply.

10.4 Effect of rectifier loads on local generators

10.4.1 *Impedance of the supply*

If the electricity supply is derived from a local generator, all the effects mentioned in Section 10.3 may be experienced, but, as the source impedance will almost certainly be higher than that of a supply system, the effects are more likely to be troublesome. There are other additional effects which occur owing to the nature of synchronous machines, and these are described in subsequent paragraphs.

The magnitude and importance of the effects described in this Section depend upon the size of the local generator. If the UPS equipment is supplied from a very large generator the effects will be negligible and the supply should be regarded as a low impedance supply system, as considered in Section 10.3. However, UPS equipments frequently have a standby supply derived from a small local generating set, and in such cases the effects may well be relevant.

10.4.2 *Effect of harmonic currents*

The simplified formula 10.7 for estimating harmonic distortion of a supply system indicates that the supply impedance determines the degree of distortion. Whilst this remains true for a supply derived from a generator, a generator is less amenable to analysis and only a brief description of the mechanisms of voltage distortion will be attempted here.

An equivalent circuit for one phase of a synchronous generator is shown in Fig. 10.6. The rotor iron and the damper windings D_d and D_q prevent rapid flux changes in the rotor, and any distortion of voltage caused by harmonic currents will be mainly due to the impedance presented by the subtransient reactance X''. This may be regarded as the stator leakage reactance, approximately equal to the mean of the direct and quadrature axis subtransient reactances.

A method is therefore available to estimate the approximate voltage distortion due to each harmonic. For a current a_n of harmonic order n the voltage across X'' will be:

$$V_n = a_n \times n \times X'' \text{ p.u.} \tag{10.10}$$

As an example, a 5th harmonic current of 0·1 p.u. passed by a machine having a mean subtransient reactance of 0·15 p.u. will produce a 5th harmonic voltage of 0·075 p.u. (7·5% of rated voltage). The total harmonic distortion can be estimated by performing this calculation for each significant harmonic current.

The distortion can be reduced by reducing the machine subtransient reactance; this can be achieved by using a larger generator frame size or, if a purpose designed machine is envisaged, by using more flux and fewer stator turns or a longer frame.

10.4.3 *Effect of commutation*

It is stated in Section 10.2 that the period of commutation of a rectifier is dependent upon the commutating reactance of its supply. If there is no rectifier transformer or series inductor the commutation period is determined entirely by the supply system reactance. A small generator will have a comparatively high

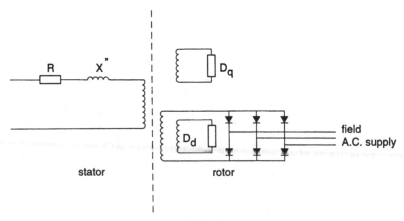

Figure 10.6   *Equivalent circuit of a generator*

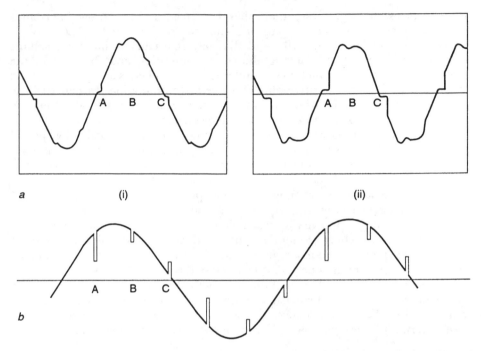

Figure 10.7 *a Examples of typical distortion caused by diode rectifier loads on a local generator*
(i) 50% rectifier load
(ii) 100% rectifier load

b Typical notching caused by thyristor loads on a local generator

reactance. It follows that, when it is supplying a rectifier, the commutation period will be extended; an extended commutation period results in a reduction of DC output voltage, and, as discussed in the following paragraphs, an increase in distortion due to notching.

When a sinusoidal three-phase generator supplies a linear load the stator currents produce a steady magnetic flux rotating with the rotor at synchronous speed. When it supplies a rectifier an entirely different set of conditions arise. The current will not be of the idealised square wave considered in Section 10.2 (Fig. 10.2) but is likely to approach a trapezoidal shape due to the extended commutation periods. Between the commutation periods the stator flux remains stationary and moves forward in discrete steps as each commutation transfers current from one phase to the next.

With a diode rectifier this results in the type of distortion illustrated in Fig. 10.7a. Owing to the extended commutation period the notching will be deeper than that experienced from a low impedance supply.

With a thyristor rectifier the operation is similar to that of diodes but the notching is deeper and is dependent upon the commutation delay as indicated by Fig. 10.7b. Owing to the irregular rotation of the stator flux the notching in the succeeding phase can be considerably deeper (and the rise in the preceding phase correspondingly less) than that experienced from a low impedance supply. Extreme cases of this type of distortion are illustrated in Fig. 10.8.

An interpretation of Figs. 10.7 and 10.8 may be useful. At time A the voltage experiences a negative notching due to commutation and the start of the conducting period. Conduction continues for 120°, and ends at time C when the voltage experiences a positive notching due to commutation. In Fig. 10.8 the notches due to commutation appear to initiate ringing of a local resonance, which adds to the appearance of extreme distortion.

Additional notching appears at B, the mid-point of the conduction period between A and C; this occurs at the time during which the other two phases are commutating. This is a phenomenon that arises when a bridge rectifier is supplied from a small generator with transient saliency. During commutation the stator flux advances rapidly to its new position, and in doing so induces reverse voltages in the stator windings. Any generator supplying a UPS equipment is likely to have a salient pole rotor and is susceptible to this effect. Theoretically the effect should be much reduced by extending the damper windings over the interpolar regions and, in practice, the effect is lessened by inter-connecting the damper windings across the interpolar gaps.

If a generator is shunt excited any notching of the stator voltage appears in the excitation circuit and can cause additional distortion. The use of a separate shunt excited AC exciter reduces the possibility, but the best arrangement is to use a permanent magnet pilot exciter followed by a main exciter.

10.4.4 *Additional rotor losses*

Harmonic stator currents drawn by a rectifier cause air gap fluxes of the same general shape as the fundamental but rotating at n times synchronous speed, where n is the harmonic number. These will induce currents in the rotor iron and windings, adding to the rotor losses and increasing its temperature.

Rectifier harmonic currents occur in pairs such as 5th, with negative rotation, and 7th with positive rotation. The 5th will induce in the rotor a negatively rotating 6th harmonic and the 7th a positively rotating 6th harmonic. The two contra-rotating 6th harmonic systems in the rotor result in an alternating flux that is stationary with respect to the rotor. The alternating flux induces damper

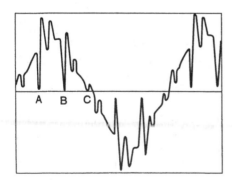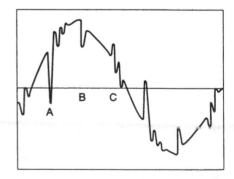

Figure 10.8   *Examples of extreme distortion caused by thyristor loads on a local generator*

winding currents that are also stationary with respect to the rotor, causing localised extra heating. Within the rotor the effect is similar to that caused by single phase or unbalanced loads. Most machines are fairly tolerant of unbalanced loads and overheating is unlikely unless the rectifier loading exceeds say 50% of the generator rating. Machines with laminated pole faces are recommended for such duties.

10.4.5 *Torque pulsations*

The resultant alternating flux mentioned in the preceding paragraph and the fundamental stator MMF are both stationary relative to the rotor. The reaction between these two MMFs results in a pulsating torque, the magnitude being dependent upon the spatial relationship between them. The spatial relationship is dependent upon the machine parameters and the commutation delay angle.

The largest torque pulsations are at six times the supply frequency, others being at twelve etc. times supply frequency. They rarely attain a significant magnitude and do not appear to be troublesome in practice. The coupling between the flywheel and the alternator will include some damping mechanism, and the effect seen by the crankshaft will usually be small compared with the torque irregularities it experiences in service.

10.4.6 *Effect on electronic devices*

A local generator is likely to incorporate electronic devices such as a tachometer and an automatic voltage regulator which may use the zero voltage crossing points as timing signals. Deep notching of the supply voltage can lead to additional zero crossings and can affect the operation of such devices. If problems are experienced a clean supply may be derived by the use of a small dedicated filter.

Automatic voltage regulators may be affected by a distorted supply; if the regulator is expected to control the fundamental component of voltage, the harmonic voltages must be removed from the sensing signal by a low pass filter. If the harmonic voltages are not removed the regulator will set to the true RMS value or the average value (including harmonic voltages) depending on its design. In cases of severe distortion this is likely to lead to a problem during operation.

10.5 Reduction of distortion due to rectifier loads

10.5.1 *Increasing the pulse number*

In Section 10.2 it is stated that the current taken by a three-phase bridge rectifier (6 pulse) includes only the 5th and 7th, 11th and 13th, 17th and 19th etc. harmonics. If the rectifier pulse number is increased to 12 the 5th and 7th harmonics are cancelled whilst the magnitudes of the higher harmonics remain unaltered. Rectifiers above the 100/150 kVA rating are frequently required to be 12-pulse type in order to reduce the generation of harmonic currents.

For the same reason the supplies to rectifiers of multi-unit UPS equipments should be phase shifted, which has an effect similar to increasing the pulse number. The supplies to two parallel units should be shifted by 30°, three units by 20° etc. For a phase shifted multi-unit redundant system, the loss of a unit will result in some additional harmonic currents being drawn from the supply.

10.5.2 *Adding an input filter*

It is possible to reduce the harmonic currents flowing through the supply system by providing a low impedance path across the rectifier power input terminals. The harmonic currents generated by the rectifier will then be shared between the supply system and the shunt path in a manner dependent upon the inverse ratio of their impedances.

The low impedance path may comprise one or more shunt connected filter circuits tuned to particular harmonics, or a simple shunt connected untuned capacitor. In either case a series inductor may be included on the supply side, which effectively increases the system impedance and thereby further reduces the harmonic current flow.

The series inductor may also serve another purpose: if properly chosen it can prevent any likelihood of the capacitor (or the net capacitance of the filter) resonating at a harmonic frequency with, for instance, the leakage reactance of a local transformer. This subject is discussed in connection with power factor correction capacitors in Section 10.3.2.

Shunt connected harmonic filters and capacitors will both take a leading fundamental frequency current which may result in the UPS operating at a leading power factor, particularly at light loads, and almost certainly if a diode rectifier is involved. At the expense of some complication it may be possible to install a capacitor bank with automatic switching, but there may well be conflict between the requirements to maintain harmonic attenuation and to control power factor.

10.6 Inverter generated harmonic voltages

10.6.1 *Inverter output waveform*

This Section applies to the types of equipment described in Chapters 5 and 6. These supply the load from an inverter output stage which produces either a square wave or a pulse width modulated output.

The UPS output is derived from two inverters which feed power into a combining transformer. The voltage waveform obtained from the inverters is barely recognisable as a sine wave but, as discussed in Section 10.1, such a waveform consists of a fundamental and numerous harmonic components. Only the fundamental component is of value in the output; the harmonic components are removed before the inverter output can be applied to its load.

In its basic form each inverter is a number of switching devices which are operated in sequence to convert a DC input to an AC output. It follows that the output has only three voltage states: the DC voltage, the DC voltage reversed and zero. It will also be apparent that the output impedance of such an arrangement is low; the current path includes only four switching devices and the DC source.

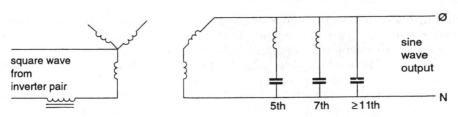

Figure 10.9 *Typical inverter output filter arrangement, one phase only shown*

10.6.2 *Attenuation of the harmonic voltages*

Because the inverter inherently has a low output impedance, it is necessary to add series impedance before the waveform can be improved. The impedance is in the form of a series inductor which, for the present purpose, could be on either side of the combining transformer. A low impedance path for inverter generated harmonic currents is provided by shunt resonant circuits down-stream of the transformer as indicated in Fig. 10.9. The primary leakage reactance of the transformer can be used to reduce the size of the inductor.

If the output is to be sinusoidal the inductor will experience a voltage equal to the difference between the inverter output and a sine wave. Such a voltage is made up of the unwanted harmonic components and, for a square wave output, would typically be of the form illustrated in Fig. 10.10. For this voltage to be induced a complex current must flow in the inductor; a path for this current is provided by the shunt filter components.

For a square wave inverter the unwanted harmonic voltages would include the 5th, 7th, 11th, 13th etc. as described in Section 10.2. Shunt paths would probably include series circuits resonating at the 5th and 7th harmonics (or one wide band circuit resonating at the 6th) and a shunt capacitor which would have a low impedance at higher harmonics.

For a pulse width modulated inverter the harmonic series would start at a higher order as determined by the pulse repetition frequency. The absence of the lower order harmonics makes the filtering easier. If the pulse repetition

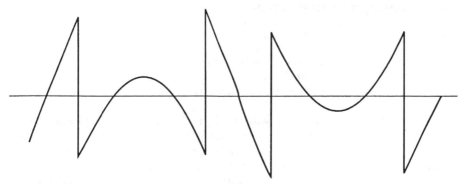

Figure 10.10 *Idealised voltage experienced by the inductor*

frequency is sufficiently high a shunt capacitor may suffice. However, if lower order harmonic currents are demanded by the load, a shunt path will be necessary; this subject is discussed in Section 10.7.

Since the shunt filter paths carry harmonic currents and do not have zero impedance the equipment output cannot be truly sinusoidal. Any shunt filter path is in parallel with the equipment load; it follows that to be effective the shunt filter paths must have harmonic impedances which are low in comparison with the load. For high power LV equipments this can be quite demanding; for example, a 500 kVA load at 415 V is equivalent to three star-connected impedances of 0·345 Ω each.

It is the lower order harmonic voltages which determine the filter parameters; the higher order harmonic voltages are of smaller magnitude and are more easily attenuated. The reactance of the series inductor is limited by the maximum acceptable transient regulation of the output voltage.

10.6.3 *Effect of a linear load on the output voltage*

It is useful to consider the effect of applying a load to the filter. A linear load is assumed, non-linear loads are introduced in Section 7.

At no load the inverter filter arrangement of Fig. 10.9 may be regarded as a potential divider, the parameters of which will vary depending upon the harmonic order under consideration. The application of a load, which will be in parallel with the shunt filter paths, will result in a reduction of the impedance seen by the inductor and thereby improve the performance of the potential divider.

A similar conclusion may be reached by considering that any harmonic voltages applied to the load will cause harmonic currents to flow, and that these currents in flowing through the inductor will cause additional voltage drops. It is somewhat of a paradox that any such reduction of distortion of the output voltage is accompanied by harmonic current flow in the load.

In practice, the effect due to a linear load will be small because the shunt filter paths will have an impedance much lower than the load. UPS loads are invariably non-linear; the concept of a linear load is introduced as it may assist readers in reaching a full understanding of the subject.

10.7 Load generated harmonics

10.7.1 *Nature of the load*

A UPS commonly supplies a load comprising data processors, computing equipment, peripherals and telecommunications equipment; each of these includes a great variety of individual loads with various distortion characteristics.

Many of the individual loads will be supplied with DC derived from switched mode power supplies, which are the main generators of harmonic currents in UPS loads. The basic switched mode power supply comprises a series arrangement of a single phase rectifier, a capacitor input filter, a chopper with associated control circuit and an output filter. The rectifier and capacitor input filter result in a peaky current flowing at the middle of each voltage half-cycle —

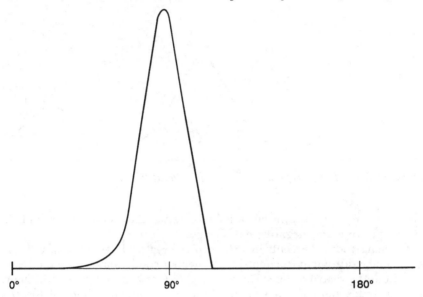

Figure 10.11 *Typical waveform of current taken by a switched mode power supply*

a waveform which has a high third harmonic content. A typical current waveform is illustrated in Fig. 10.11.

The waveform of Fig. 10.12 is a useful approximation to that of Fig. 10.11; it is made up of the following harmonic components:

Harmonic order n	1	3	5	7	9	11	13
Magnitude %	100	85	60	30	18	12	8

The total load current demanded from a UPS equipment is likely to include a high proportion of various odd harmonics, particularly the third. It is unlikely

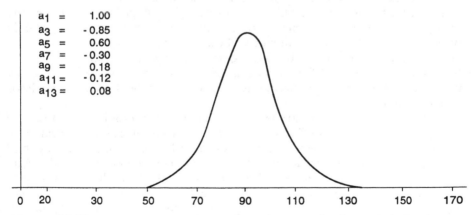

$a_1 = 1.00$
$a_3 = -0.85$
$a_5 = 0.60$
$a_7 = -0.30$
$a_9 = 0.18$
$a_{11} = -0.12$
$a_{13} = 0.08$

Figure 10.12 *Approximation to the waveform of current taken by a switched mode power supply*

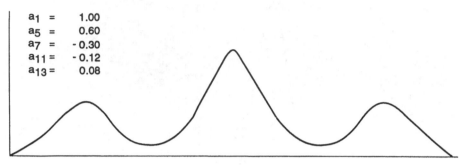

$$
\begin{aligned}
a_1 &= 1.00 \\
a_5 &= 0.60 \\
a_7 &= -0.30 \\
a_{11} &= -0.12 \\
a_{13} &= 0.08
\end{aligned}
$$

Figure 10.13   *The waveform of Fig. 10.12 without triplen harmonics*

to include any significant even harmonics as they indicate an unbalance between positive and negative half cycles.

Triplen harmonic currents in a three-phase system have zero phase sequence and become additive in the neutral; the neutral conductors within the UPS and throughout the distribution system must therefore be rated for an RMS current which includes both the design fundamental component and all the triplen harmonic components generated by the load.

10.7.2 *Effect of a non-linear load on UPS equipments with inverter generated output*

The types of UPS equipment under consideration are described in Chapters 5 and 6. Such equipments invariably incorporate a double wound combining and output transformer which provides a neutral and a three-phase output for the load. The transformer is also able to provide a low impedance path for load generated balanced triplen harmonic currents; the low zero sequence impedance can be achieved either by arranging the secondary (output) winding as an inter-connected star with tightly coupled pairs of windings, or by providing a tertiary delta tightly coupled to the secondary. It is now apparent why the inductor of Fig. 10.9 is shown upstream of the combining transformer.

The transformer is not able to provide a low impedance path for any unbalanced triplen components; these will appear in the primary, and in passing through the series inductor will cause voltage distortion in two phases. So far as possible, therefore, load generated triplen harmonics should be balanced. The effect of unbalanced triplen currents can be reduced by providing a shunt resonant path, but the low order of the third harmonic leads to large components.

If the triplen harmonics are removed from the waveform of Fig. 10.12, it is changed to that shown in Fig. 10.13. The new waveform is far from ideal but it is a considerable improvement; the peak current has been reduced by one third and the period of conduction has been extended to the complete half cycle. With the triplen harmonics removed there remain only the 5th, 7th, 11th, 13th etc., which are also the harmonics characteristic of three-phase rectifiers and of quasi square wave inverters. A low impedance shunt path must be provided for these harmonic currents, otherwise there will be severe distortion of the UPS output voltage.

It may be of interest to the reader to compare the waveform of Fig. 10.13 with that of Fig. 10.1*b*. In both cases the removal of triplen harmonics results in a waveform characterised by three peaks.

Quasi square wave inverters will, as explained in Section 10.6, require shunt filter paths for harmonic orders down to the 5th; a low impedance shunt path will therefore exist for all odd harmonic currents generated by the load.

Pulse width modulated inverters will require shunt filter paths for higher order harmonic currents only. However, it follows from the foregoing that a low impedance shunt path must be provided for each load generated harmonic current.

The shunt filter paths will carry the harmonic currents generated by both the inverter and the load as indicated in Fig. 10.14. The paths must be rated to carry the resulting current which will depend upon the magnitudes of the currents, which can be estimated, and their phase relationship, which is not easily predicted. In practice, the rating of the shunt filter components should be adequate for the arithmetic sum of the inverter and load generated harmonic currents, in addition to any fundamental frequency current which will be drawn.

10.7.3 *Effect of a non-linear load on UPS equipments with machine generated output*

The types of UPS equipment under consideration are described in Chapters 3 and 4. The effects of distorted currents on the performance of synchronous generators are discussed in Section 10.4, and the same comments are applicable here, the difference being that the generator is now supplying the UPS load

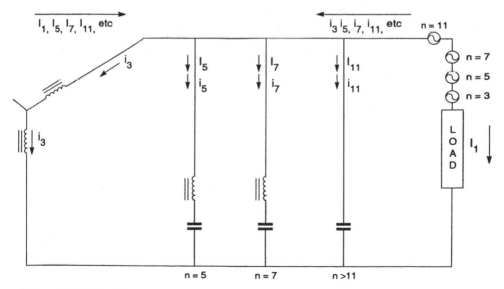

Figure 10.14   *Harmonic current paths in the inverter output circuit*

whereas in Section 10.4 it was considered to be supplying the UPS input rectifier.

A rotary generator used at the output of a UPS equipment is dedicated to the UPS load and will probably experience a severely distorted current. It will operate without the advantage of the mixed load that might be experienced by a standby generator supplying power to the UPS equipment as well as other essential loads.

The load is unlikely to include any significant thyristor rectifier equipment, and the effects due to notching are unlikely to occur. The effect of interest to the user will therefore be distortion due to harmonic currents; the manner in which this arises is discussed in Section 10.4.2. Other effects, such as additional rotor losses and pulsating torque, are of interest to the manufacturer and should be taken into account at the design stage.

It is apparent from what has been written in Section 10.7.1 that the load may well demand significant third and other triplen harmonic currents. The inverter type of UPS equipment provides a low impedance path for these through the combining transformer, but when the output is taken from a machine it is usual to pass the total current through the machine windings. If the harmonic currents are likely to cause unacceptable voltage distortion the distortion can be reduced by interposing a delta/star (or star/interstar) transformer between the generator and the load. An interstar wound static balancer would achieve the same result but the load must use the balancer neutral, leaving the generator neutral to float. The rating of the UPS installation must be adequate to supply the transformer losses and must have sufficient voltage in reserve to allow for the voltage drop within the transformer. The transformer neutral must be adequately rated and a return path for the triplen harmonic components of the magnetic flux may be necessary.

Chapter 11

UPS reliability

P. Smart

11.1 Reliability

Before investigating the degree of reliability of a system or the factors which affect reliability, it is necessary to firstly define the term 'reliability' and the system contained within the meaning of UPS.

Uninterrupted power supply, or UPS as it is more commonly referred to, usually means the battery and static or rotary module(s) are provided to ensure that a continuous supply is maintained for a predetermined time period. The total uninterrupted power concept, however, embraces not only the UPS but also the generators, switchgear and power distribution units which combine to give a supply with almost indefinite time limitations, and particularly in large installations.

The term 'reliability' seems to have a different interpretation and requirement depending on whether one is a manufacturer or a client. Much has been said about reliability of UPS equipment in various general articles and in manufacturers' technical publications, but this is usually subjective and comparative rather than objective and constructive. One dictionary definition of 'reliability' is 'that which can be trusted' but again, depending on our own particular requirements, what is reliable for one may not be reliable for another. For example, if a business operates for ten consecutive hours each day in a five day week, then providing a system works without interruption within that period it is considered reliable. However, another business working 24 hours per day, 7 days a week, 52 weeks per year may classify the system as reliable only if there are no interruptions at all. One failure per month may not be of great concern to some, whereas one failure a year may be catastrophic to another.

The general purpose of a UPS system is twofold: one to give continuity of supply, and the other to provide a clean supply (consistent within set limits of variation for voltage and frequency). Again, depending on the client, either may be the more important and this influences the interpretation placed on the term 'reliable'.

Manufacturers, however, tend to have a much clearer interpretation and are prepared to quote figures indicating their own 'mean time between failure' (MTBF) assessments. The MTBF figures normally relate to the UPS module and not to the whole system and are calculated under ideal conditions. The user, unfortunately, can only use these figures as a guide because business operations are normally dependent on the total uninterrupted power concept

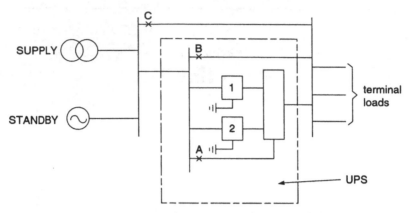

Figure 11.1   *Simplified diagram of UPS network*

and this is made up of component parts all of which have their own reliability factor and the conditions are variable. A further problem in assessing reliability is that one component may have a major impact on another.

A total UPS system can, as in many other engineering systems, be made more reliable by the introduction of parallel redundant modules and additional battery back-up, but these have a considerable impact on the cost of the installation in terms of space and money.

Perhaps to assist in the understanding of what, for the purposes of this chapter, is termed the UPS network and its reliability, the simplified diagram of Fig. 11.1 will identify components and their relationships. Note that this section considers a central supply arrangement only. Modular units will be discussed at the end of the chapter.

The term 'supply' has been used to identify the prime 'source', usually the local electricity supply utility, while the generator has been taken as the standby supply; however, this may be reversed. The UPS block may be taken as either a single- or multi-module unit and for the larger capacity system an automatic by-pass is normally provided which will in the event of the module(s) fault provide a by-pass to the load without power interruption. When operating in this mode, supply is directly to the load terminals and is, consequently, subject to any mains disturbance. UPS may be of any style: rotary, transistor, inverter etc. Batteries suitably sized for the specified autonomy are either housed within the same area as the UPS or separately. It can be seen that a client whose prime object is to have a clean supply will view as unreliable a system which gives continuity of supply via the by-pass even if the modules themselves are subject to a number of faults. This view may change, however, if one is more concerned with a 'continuous' supply.

11.2 Factors affecting reliability

Like any other item of plant, the reliability of a UPS system is dependent on a number of factors not all of which are the responsibility of the manufacturer. The particular points in question which have a substantial bearing on the UPS operation are:

(*a*) System design and space planning
(*b*) Installation and environment
(*c*) Maintenance and operation

Let us consider each of these individually:

11.2.1 *System design and space planning*

The initial system design can be expected to be undertaken by a combination of, or a single member from: consultant, client, manufacturer or contractor. It is far too easy to have a UPS installation carried out without sufficient planning having been undertaken. This is particularly so over the last few years when the UPS market has taken off and the number of suppliers has multiplied enormously, without the same degree of installer and user knowledge.

In any design the system must be right for the purpose. From the outset one must determine:

- Prime purpose
- Operating hours
- Type
- Capacity
- Battery
- Systems
- The cost.

If the initial design is wrong in size, then modules may frequently be overloaded. If it is wrong in its interpretation of use, then it may be that problems occur at the load terminals outside the units. Problems frequently encountered are harmonics, voltage spikes and phase separation.

Looking specifically at these points, we should recognise that our failure to design correctly can be costly in terms of both money and time lost. What is the prime purpose of installing UPS equipment: is it to protect our business or the equipment we use? If it is the former then frequent use of the auto by-pass may be acceptable and our reliance on the actual UPS conditioning may not be the dictating factor. However, if our equipment is sensitive to supply disturbances, then we are likely to spend more time ensuring it works correctly. This means that we all have to know the 'Business'. The supply parameters set by many equipment suppliers such as IBM, DEC etc. now create a more positive line of thinking in that warranties may be affected by the introduction of, or lack of, controlled supply and environment.

Next we need to consider the time for which we expect the equipment to be operating, as the longer we wish the system to be serviceable the more we will have to increase the reliability factor. The introduction of recent legislation is helping to persuade businesses that they must allow down time for maintenance, and this gives a slot for maintenance and testing of the system. Previous chapters have discussed the various types of UPS and one needs to determine which particular type suits the application in which we are interested. Many factors can influence our decision, but one often overlooked is the emergency response and call-out times offered by the manufacturer. If the response is delayed this can cause considerable inconvenience and possible business loss.

The capacity of the module must be determined taking into account the load and the amount of redundancy which is required. A multi-module system connected in a parallel redundant mode not only allows for redundancy but for routine maintenance as well as increasing the reliability of the system.

The capacity of the system, both immediate and longer term, needs to be considered at the outset if one is to maximise the cost invested. The obvious alternatives are to:

(i) Leave space for future modules and batteries
(ii) Provide adequate batteries for the final system, giving increased autonomy in the short-term
(iii) Plan for additional stand-alone systems.

As UPS systems vary in capital cost depending on the size, type and system, it is essential to make the correct decision at the outset, and the design criteria is therefore critical.

11.2.2 *Installation and environment*

The installation and environment in which the equipment is expected to operate have a far greater impact on the equipment than one would generally expect. In most instances, the manufacturers will ensure that the items are correctly packed and transported, but, at the time of off-loading and installation, the same care is not always taken. All electromechanical equipment such as UPS and generators need to breathe, and therefore adequate air movement and cooling is required. One not only has to provide adequate space for this but also allow for ease of access for operation and maintenance.

If this is the case, then the installation should be safely carried out. It is always preferable to have the manufacturer or authorised supplier supervising the off-loading and site placing so that no warranties are affected. In the light of current regulations (Electricity at Work Regulations), the provision of adequate operation and maintenance space and accessibility will be the responsibility of client, consultant, architect, manufacturer and contractor. Compliance with this regulation will greatly assist in leaving space for air movement but may well restrict the capacity of the system which can be safely contained in the area; this, in turn, may lead us back to design criteria — central units versus local stand-alone units.

It is important if maximum reliability is to be maintained that all associated cables and switchgear are installed correctly and with the correct capacity. While the UPS supplier is usually responsible for ensuring the switches are of correct size it is advisable to check that (*a*) the auto by-pass (*b*) by-pass and (*c*) full maintenance by-pass switches are rated for the full system load and not the single module or design capacity. Similarly the cables should be fully rated and installed for full system load.

With regard to the environment in which the equipment is placed, this has, or can have, an effect on the operation of equipment and its autonomy. UPS modules are generally sturdy units with inbuilt fans to pull air across the components so that the unit runs at a steady temperature. The operating temperatures do vary from manufacturer to manufacturer but are generally of the order of 25°C with an upper limit of 45°C. When this is reached, the modules are shut down to by-pass. Obviously, if this occurs while the UPS is functioning from the battery, supply to load terminals is lost. To overcome this, it is essential that the air within the plant room is 'conditioned'. A totally enclosed room with no cooling will soon reach the cut-off temperature. If the room is air-conditioned, then the supply to the air-conditioning unit must be fed from the standby supply so that it is restored with minimum delay.

If the design is such that a 15 or 30 min autonomy is required, then one must check that, in the event of a simultaneous fault on both main and standby supplies, the temperature rise within the plant room does not increase at such a rate that the UPS shuts down on high temperature before the autonomy has been exceeded. This is particularly important in situations where the design allows for longer autonomy at less than full load as the rate of temperature rise does not decrease to the same degree as the load is reduced. In a multi-module set up one must avoid hot spots in the room in order to ensure that it does not overheat and trip out with consequent cascading of the others.

The battery location varies from site to site as do the conditions in which they are found. As with most equipment, batteries work to their maximum capacity between certain temperatures, and constant variations cause loss of power and deterioration. Ideally, the battery should be in a temperature of about 20°C. The life expectancy decreases by some 25% for each 15 deg C above 20°C, so the reliability can be seen to be partly determined by the environment.

The lower the temperature the more likely it is that the cell capacity would have to be increased, and the higher the temperature the more likely that additional maintenance will be needed and the lower the battery life.

11.2.3 *Maintenance and operation*

As with all equipment, it is necessary to carry out regular and detailed maintenance of the UPS modules, batteries and environmental control equipment. Unfortunately, this vital requirement takes time, costs money and may mean disruption to the end user; hence service periods are often extended or overlooked. Some may argue that the system is best left until it goes wrong, but, in my view, this is negative in that it is probable that the equipment will fail just as it is needed.

Provided that thought has been put into the design, it should be possible to maintain most, if not all, of the system, without there being interference to the supply at the load terminals. Maintenance schedules are prepared and carried out by manufacturers and major suppliers, and during the maintenance it is often possible to highlight any changes in the system such as increased load, large out of balance, and increases in temperature both within the module and the environment. The point settings can also be checked and varied according to requirements. This is important because, as terminal equipment is developed, so the tolerances of voltage and frequency are often compressed, and if not monitored the equipment can fail owing to them not being kept within the limits. It is also important to check the system autonomy at least once per year in order to ensure both that the system still maintains its design parameters and also that it assists in keeping the battery at maximum capacity.

11.3 Stand-alone systems

The foregoing has been particularly relevant to central systems, but much of what has been said also relates to stand-alone systems and it should be remembered that these systems too need maintenance.

A stand-alone system is one which, where the module is usually static, battery and controls come as a one-piece item and are free to serve either a single PC or a network, depending on the size of the module. Stand-alone units come with either single- or three-phase operation and have stated mean time between failures of 80–100 000 hours, although the recently introduced range of Ferrups and Microferrup systems give some 200 000 hours. It should be noted that the battery is not included in these figures.

Stand-alone systems are generally designed for use in a normal office environment but they can be used wherever the need for UPS arises. However, in all instances it is important to ensure that the environment complies with the manufacturer's guidelines or reliability will be reduced and the warranties will be nullified.

11.4 Conclusion

In conclusion it is important to recognise the impact of UPS on the business and to ensure its service and reliability by monitoring its maintenance and environment irrespective of whether the system be of the stand-alone or central system format. While it is impossible to make recommendations as to which systems are the best when it comes to reliability it can certainly be concluded that most systems are tested and designed for reliable service, but this can either be improved or impaired by the care and attention given to the total installation approach.

Chapter 12

The specification

A. C. KING

12.1 Introduction

Although UPS equipments have been used increasingly since the late 1960s there is a dearth of formal specifications and guidance material. At the time of going to press, there are two IEC specifications relating to UPS equipments which are appropriate to mention here: IEC 146–4, 'Semiconductor converters – method of specifying performance', was published in 1986 and includes an extensive glossary of terms and definitions, defines the service conditions and lists those that should be identified by the purchaser. Tests are specified, it being a matter for agreement as to which tests are undertaken at works and which at site; IEC 146-5, 'Method of specifying switches for uninterruptible power systems' was published in 1988 and defines the many types of power switching devices used within UPS equipments.

It may seem a trite statement, but before a specification for a UPS equipment can be written, it is necessary to consider carefully what is expected of the equipment. It is the author's opinion that a weak original specification is frequently responsible for the failure of an equipment to perform as expected. Important items are sometimes omitted from the specification either from oversight or because they are too difficult to express; it is the purpose of this Chapter to assist purchasers to decide their functional requirements, and to include them in a specification. It is written in the form of an aide memoire; additional information (in italic print) is given where this is appropriate.

The list is divided into the following Sections:

12.2 Location of equipment
12.3 Type of equipment
12.4 Earthing the neutral of the UPS output
12.5 Input power — ratings and quality
12.6 Type of load
12.7 Output power — ratings and quality
12.8 The battery and battery charger (if applicable)
12.9 Service conditions
12.10 Installation requirements

12.2 Location of equipment

Is the equipment to be installed in an 'office' environment, or in a 'plant room' environment?

For large equipments there is probably no choice.

12.3 Type of equipment

Single or parallel units, with or without redundancy?

There is no advantage in specifying parallel units unless any of the following apply:

Redundancy is specified
The load is such that it can run on reduced power on the loss of a unit.
Rectifiers are phase shifted to reduce the harmonic currents taken from the mains
Load growth is anticipated

Is an automatic by-pass required, to normal supply or to standby supply?

A by-pass is only possible if the input and output frequencies are identical; if the input and output voltages differ, an interposing transformer will be necessary.

An automatic by-pass is only available whilst the by-pass supply is within the specified tolerances, and the UPS is able to run in synchronism with it. It follows that an automatic by-pass is of value only if the by-pass supply is of adequate quality to supply the load; a standby supply in particular may suffer from quite large voltage and frequency excursions.

Is a hand operated maintenance by-pass required? Should it be internal (within the UPS equipment), or external so that the UPS equipment can be made totally dead?

An external by-pass may be considered desirable for important installations. The advantage of an external make-before-break by-pass circuit, entirely separate from the UPS equipment, will be apparent to any reader who has experienced a failure requiring extensive remedial work on the UPS equipment.

12.4 Earthing the neutral of the UPS output

How are the neutrals of the UPS output and the by-pass supplies to be earthed?

The question relates only to the type of UPS equipment which provides electrical separation between the input and output supplies. National regulations and codes of practice usually require that the UPS output shall be bonded directly to earth, and that any exposed conductive metalwork shall be bonded to earth. It is not appropriate here to dwell on any particular regulations, but some general advice is offered:

(*a*) UPS installations without a by-pass

Where the installation is supplied from a local distribution transformer it is good practice to connect to earth at a point electrically close to the earthing conductor (which connects to the electrode system). This will be a 'clean' earth and minimises interference due to rogue fault currents flowing in the neutral or in the protective conductors. However, if a by-pass supply is to be provided this may not be practicable for the reasons given in (b) below.

Where the installation is supplied at low voltage and there is no local distribution transformer the earthing conductor is unlikely to be accessible and it is necessary to use the installation earth (or the supply neutral) as the earth. The connection to the neutral should be at the intake position and not within the installation. Provided that the metalwork within the building is properly bonded to earth, the use of the supply neutral as the earth is unlikely to result in spikes or other interference being injected into the UPS output. Note that it may not be permissible to bond the supply neutral to an earth electrode within the installation.

(*b*) UPS installations with a by-pass

Where a by-pass supply is available as an alternative to the UPS output, it is necessary for proper operation for the neutrals of the UPS output and the by-pass supply to be at the same potential; it is therefore usual to use three-pole switching and a solid neutral in the by-pass and to solidly connect the UPS output neutral to the by-pass neutral. The UPS output neutral relies on the by-pass neutral for its earth. The by-pass supply should therefore be derived from a point close to the intake; this will be the cleanest supply available.

It rarely occurs but it is possible for the by-pass neutral to introduce interference into the UPS output. Ideally the interference should be identified and isolated or suppressed, but if this is not practicable and the risk is not acceptable, the by-pass supply may be separated from the UPS output by a double wound transformer. The UPS output neutral and the by-pass neutral can then be connected together and earthed as described in (a) above.

The transformer will introduce voltage regulation into the by-pass supply. It must be designed to accept the UPS load including its triplen harmonics; the core may require an additional limb as a flux return path.

12.5 Input power

Normal supply voltage, frequency, phases, 2, 3 or 4 wire
  Voltage and frequency tolerances + %, − %.
  Prospective short circuit current in kA.

Is a standby supply available or envisaged? If so the following information should be provided:

  Voltage and frequency tolerances + %, − %
  Maximum acceptable step load
  Diesel or gas turbine driven
  kW rating of the generating set
  Is it dedicated to the UPS or shared with other building services?

A naturally aspirated diesel engine or a single shaft gas turbine will normally accept a 100% step load. A supercharged set will have a lower limit, down to perhaps 40% depending upon the degree of supercharging.

A two shaft gas turbine likewise has a poor step load performance but is not much used for standby generating purposes.

Where the UPS equipment incorporates a thyristor rectifier, the step load on restoration of the supply can be reduced by providing a 'walk-in' feature.

Is there any restriction regarding the harmonic currents that may be taken from the normal supply or the standby supply?

The local supply authority will invoke Engineering Recommendation G5/3 (see Section 10.3) or some other equivalent code. If the installation takes power at a high voltage, the point of common coupling should be regarded as the UPS input terminals, thus preserving the quality of the internally distributed power. Any standby supply will be vulnerable to distortion; if it supplies loads other than the UPS, the results of distortion on those other loads should be considered.

If the standby set is dedicated to the UPS the degree of distortion will be high but probably not important.

If the UPS equipment is provided with a by-pass there may be occasions when the computer load will be connected directly to the supply (either the normal or the standby supply).

12.6 Type of load

RMS value of load current (including all harmonics).

Rating of load in kW (using fundamental current component) and kVA (using RMS value of current).

Power factor of fundamental component of current.

Harmonic content of load, or distortion factor, μ.

Peak factor of load current.

Phase unbalance of load current.
   Information on probable phase unbalance will be difficult to obtain, it is suggested that a figure of 0·25 is used as defined in IEC 146–4.

Type of load, e.g. data processing, peripherals, process control, motors, rectifiers.
   A computer type number may provide useful information to the UPS manufacturer.

Start-up procedure and maximum expected inrush current.
   Data processing equipment with discharged capacitors can take heavy switch-on surges. The effect can be reduced by sequence starting of parts of the load.
   Data on the load are often difficult to obtain, but some guidance must be given to the manufacturer if the performance is to be guaranteed. UPS manufacturers can obtain much useful information if given details of the computer to be supplied.
   If assumptions have to be made, the following figures are offered as representing a load which includes 75% of switched mode power supplies:

   cos φ of fundamental   = 0·9
   Distortion factor μ   = 0·76
   Ratio kW/kVA   = 0·68 (from cos φ and μ)
   Total harmonic distortion = 0·85
   Peak factor   = 2·6

12.7 Output power

Minimum rating of equipment in kVA
   This will be derived from the load kVA, increased by factors to take account of:
   Peak factor or distortion factor of the load
   Any anticipated future load growth
   Rounding upwards to a convenient rating

Output voltage, frequency, phases, 2, 3 or 4 wire.

Maximum acceptable voltage and frequency tolerances under steady-state load conditions:

Voltage − %, + %: Frequency − %, + %.

Typical figures would be ±2% of voltage and ±1% of frequency

Phase angle tolerance e.g. 120°±1% for balanced loads

Maximum acceptable voltage and frequency swings under dynamic load conditions:

  Due to a 50% step load application
  Due to a 50% step load removal
  Due to 100% step load removal

The 50% load step is taken as a typical figure and is probably larger that any step likely to be encountered in practice. The voltage and frequency after 100% load removal might be of interest if there is a small residual load such as an indicator panel which may be damaged by any resulting overvoltage. The output frequencies of UPS equipments described in this book are independent of the load; any changes in frequency will usually be the result of the need to remain in synchronism with the by-pass supply.
  Typical figures for voltage would be:
  For 50% load steps: − 8% and + 10%
  For 100% load step: + 15% of voltage
  Recovery time: 100 ms

Maximum acceptable harmonic voltage distortion when supplying the type of load described in Section 12.4.
  IEC 146–4 and most manufacturers' technical literature refer only to the harmonic distortion under linear load conditions. It would be useful if there was a standard non-linear load which could be invoked for specifying purposes. In the absence of such a standard, the required performance can be specified by referring to the defined load. Typical acceptable figures might be for the total harmonic distortion not to exceed 10% and for no single harmonic to exceed 5% of the fundamental. All the requirements relating to the output power should be determined by the computer manufacturer.
  Computer manufacturers sometimes define the range of tolerable disturbances by means of two curves which plot overvoltage and undervoltage against time. The upper curve represents the computer overvoltage withstand limit, and the lower curve the computer undervoltage ride through limit. IEEE Standard 466 includes such voltage tolerance curves. If the computer voltage tolerance curves are available, they may be compared with the disturbances predicted from the UPS, and from any alternative by-pass supply, due to load changes and fault clearance.

Performance required during a fault condition downstream of the UPS equipment.
  A fault downstream of the UPS equipment will be cleared by the operation of a fuse or circuit breaker. The UPS equipment or the by-pass supply (which may also be the standby supply) must have the capacity to operate the largest downstream protective device.

12.8 The battery and battery charger

This Section is not relevant to equipments such as those described in Chapter 4 which do not use a battery energy store.

12.8.1 *The battery*

Type of battery

Will the battery be installed alongside the UPS equipment or in a battery room?

The choice is between sealed recombination lead–acid (which can be installed alongside the UPS equipment), traditional lead–acid or nickel–cadmium alkaline (either of which will usually be installed in a battery room).

If parallel UPS units are specified, is one common battery acceptable, or is one required for each unit?

Although batteries are very reliable, it is not unknown for a battery bank to be left out of circuit, or for a link to be omitted!

Period of autonomy when supplying the specified kW load.

The period is usually 10 or 15 mins; it should provide time to start and connect any standby supply, and to shut down the computer in a tidy state if there is no standby supply or if the standby supply fails. At the end of the battery's useful life, the capacity will probably have decreased to 80% of the initial capacity; if the period of autonomy is critical it should be increased by 25% to allow for degradation. The battery is an expensive part of the UPS equipment and the capacity should not be larger than the importance of the installation justifies.

12.8.2 *The battery charger*

The charging regime should be current limited constant voltage in accordance with the battery maker's recommendations.

The charging regime used for UPS duty is invariably current limited, constant voltage.

The ripple current should not exceed the limit recommended by the battery maker.

Ripple current reduces the life of a battery; in the absence of specific data a typical limit would be for the ripple current in amperes not to exceed 7% of the battery 3 hour ampere-hour capacity during recharge or float operation whilst supplying load.

Charger failure should initiate an alarm.

Period for recharging after a period of discharge.

For efficient charging and a long battery life the charging current should be limited to a figure related to the ampere hour capacity of the cells. If a short recharge time is specified the ampere hour capacity of the battery may have to be increased merely to accept the required charging current. Under a constant voltage charging regime a battery's rate of charge acquisition is asymptotic; it follows that a very long time is needed to reach the 'fully charged' condition. The specification should not therefore require the 'fully charged' condition to be attained during commissioning tests. The following duty cycle is suggested (starting with a 'fully charged' battery):

> *Discharge at rated UPS power for 15 mins (or alternative period of autonomy)*
> *Recharge for a period of 3 hours*

Discharge at rated UPS power for a period equal to 80% of the period of autonomy specified.

If parallel units with redundancy and separate batteries are specified, this duty cycle should be possible with any redundant units out of action.

12.9 Service conditions

Minimum ambient air temperature for UK installations suggests 5°C for the battery and 0°C for other parts.

Maximum ambient air temperature for UK installations suggests 30°C for the battery and 40°C for other parts, both with a maximum relative humidity of 80%.

A minimum temperature of 5°C is given for the battery because a low temperature adversely affects performance.

Altitude above sea level.

Is the location clean, or is there a likelihood of dust, chemical fumes or other deleterious substances affecting the equipment?

The degree of protection offered on standard equipments will probably be IP21. In a dusty environment filtration will be necessary; in a polluted environment a closed ventilation system may be necessary.

Will the equipment be subject to vibration?

12.10 Installation requirements

Type of cables to be used, entering from above or below?

Is noise an important consideration?

This could be important in or near an office environment. It is, however, not easy to specify requirements or to prove compliance after installation.

Is electromagnetic radiation from the UPS equipment likely to be a nuisance?

If so, the German specification VDE 0875 may prove useful in specifying any limitation.

Is electromagnetic compatibility important?

This could be important, for instance, near radio transmitting equipment, or even near local radio-controlled equipment. Microprocessors within the UPS equipment could be affected by the radiation. At the time of going to press the UK legislation relating to European EMC requirements is not in place but is expected to become law during 1992.

Is a remote status or alarm indicating panel required? If so is it to have an office compatible appearance?

Is there unimpeded access to the proposed equipment location? Are there any dimensional or weight limitations?

Index

Printed in the USA
CPSIA information can be obtained
at www.ICGtesting.com
JSHW011510221024
72173JS00005B/1267

9 780863 412639